ELECTROPOLLUTION

D0686117

About the author

Roger Coghill, a Cambridge graduate in biology specializing in brain function, has spent many years researching into the mechanisms of electropollution and its effects on living creatures, and has become an internationally acknowledged authority in this field. He runs a specialist consultancy in electromagnetic pollution monitoring and control for factories, offices and homes, and is also conducting research into electromagnetic implications in cot deaths and leukaemia. He has had published many articles and papers in scientific journals, national and local newspapers, is a frequent broadcaster on radio and TV, and when not away lecturing at home and abroad, he lives in an old cottage at the edge of a Welsh National Park.

To Chris and Lizzie, whose company uplifts my soul.

ELECTRO POLLUTION

How to protect yourself against it

ROGER COGHILL

THORSONS PUBLISHING GROUP

First published 1990

British Library Cataloguing in Publication Data

Coghill, Roger
  Electropollution : how to protect yourself against it.
  1. Man. Health. Effects of electromagnetic radiation
  pollutants. Environment. Pollutants. Electromagnetic
  radiation. Effects on health of man
  I. Title
  613.1

ISBN 0 7225 2307 6

*Published by Thorsons Publishers Limited, Wellingborough,
Northamptonshire NN8 2RQ, England*

Printed in Great Britain by MacKays of Chatham, Kent
Typesetting by MJL Limited, Hitchin, Hertfordshire

10 9 8 7 6 5 4 3 2 1

Contents

The Electromagnetic Spectrum
Electromagnetic Communications and Health

* In the text the first figure in brackets refers to the chapter (e.g. Chapter One, reference one is 1.1) and the other figures to the references at the back of the book (e.g. 68;153 are references 68 and 153. The references themselves are also back-referenced to chapters via the figures after the name(s) heading each reference.

Introduction

The origin of diseases should be obvious. . . for if there is any
unnatural formation or change of place in the elements, parts
that were cool become moist, what was light becomes heavy, and
every kind of change takes place.

The only way, in fact, in which good health can be maintained
is for replacement and waste to be uniform, similar, and on the
same scale; any trespass beyond the limits in the process will
give rise to all kinds of change, and to endless disease and
deteriorations.

Plato of Athens, *Timaeus*, 44, 82.

This book is about electropollution, the damage caused to living creatures by electromagnetic energy.

Human beings have been on this planet for some three million years. Only in the last few decades have we ever used electromagnetic energy for lighting, heating and powering our
homes, factories and offices; before 1900 nearly everyone used
gas, paraffin and solid fuels for that purpose. Moreover, we now
exploit other properties of the same electromagnetism–namely
its ability to cause action at a distance–to communicate from
one end of the world to the other, and even beyond, through
space, to distant planets or spacecraft.

It would be naive to think that nature ever bestows a riskless
benefit. There may well be a long-term biological price to pay
for the profligate use of this newly discovered power. This was
certainly the case with x-rays, also discovered less than a
hundred years ago; it took thirty years for scientists to realize
that x-rays were dangerous and to impose permitted exposure
limits. During that painful period many early researchers lost
limb and life through unwitting exposure to them. The history

of x-radiation thereafter has seen constant downward revision
of the maximum permitted exposure levels.

I can remember as a boy being able to enter a shoe shop and
put my feet into an x-ray machine to see the bones in my feet
as often as any cheeky lad might wish to, which today would
be unthinkable. We have learned to treat radiation with the
greatest respect. Even the obsession with sunworship has
diminished, as people have found out how easily it causes skin
cancers; our sun is still the most important source of electromagnetic energy we may ever encounter. Finally, current scientific
concern about the ozone layer is really concern that if it is lost
then ultraviolet rays from the sun will extinguish organic life
on earth.

The aim of this book is to tell people about the hazards of
electropollution in its many insidious forms, and suggest means
of protection.

Scientists call x-rays 'ionizing' because they are powerful
enough to knock electrons off the atoms they encounter. But
both these and the non-ionizing electromagnetic energy which
we use in our everyday lives are really part of one continuum,
the electromagnetic spectrum, and there is no real difference
between the two sorts of energy. An increasing number of scientists believe that electromagnetic energy of the non-ionizing kind
poses the same hazards as ionizing energy, but over a longer
timescale. This book summarizes the evidence and suggests
practical ways in which such hazards can be avoided or at least
mitigated.

The central concept advocated herein may well cause a sensation in the medical world. Some will brand it as complete
heresy–a fitter candidate for burning than the works of Paracelsus
or Galileo Galilei–and will call for recantation by its author.
Another writer, Rupert Sheldrake, who, dissatisfied with the
unsolved questions of biology, offered similar radical ideas concerning morphic resonance, was certainly reviled by orthodoxy.

Others may hail the idea as the most important conceptual
advance in our understanding of biological systems since William Harvey discovered the circulation of blood in our arteries
and veins. Both will be wrong. For although this book offers
to the reader a simple hypothesis to explain how we are affected
by electropollution, it will still require teams of specialists round
the world to test the predictions it makes, to formulate the
mathematical constructs which will give it establishment accep-
tance, and to define and perform the delicate experiments which

will confirm its mechanisms. In short, others must carry out the humdrum routines of the scientific community in modifying, debating, refining, and applying the discovery for the lasting benefit of all creatures competing for life and happiness on our planet. To that community of physicists, chemists, microbiologists, histologists, geneticists, cytologists, immunologists, and medical researchers, this work is respectfully dedicated.

The main purpose of this book is to tackle the invisible threats electropollution poses to our newly-identified electromagnetic mechanisms of life. Armed with the insight into how many modern diseases owe their origin to electropollution, the task, as you will see in the chapters which follow, becomes much easier.

Roger Coghill
Ker Menez
Lower Race, Gwent. 1990.

Acknowledgements

Many people have helped in the compilation of this book–too many to be listed individually–and I thank them all.

Nevertheless, I would like to single out for special gratitude Simon Best, Alisdair Phillips, and Leslie Hawkins, who gave interested and thoughtful consideration to the contents via a multitude of conversations. My thanks also to Manny Patel and all at the Life Cancer Foundation, who listened so appreciatively, and to Laurence Bloom, who did more than listen. To Eve Pollard of *The Sunday Mirror*, and Joan Lestor, M.P. who stood by their convictions and believed. To Rupert Sheldrake, Bob Liburdy of Lawrence Berkeley, Carl Blackman of E.P.A., and Paul Brodeur, Andy Marino, Bob Becker, Lyall Watson and Pythagoras of Samos, who all know more about this thing than I do.

I am grateful for permission to reproduce material and quotations from *The New Yorker*, from The Presence of the Past, courtesy of Collins Publishers, from various N.R.P.B. and H.M.S.O. publications and from many other related textbooks and publications listed in the references.

I also owe a debt of gratitude to the pioneers of the medical and biological studies which preluded my understanding of how organic life really works–to Albert Abrams, Georges Lakhovsky, Albert Szwent-Georgyi, Samuel Hahnemann, Freiherr Gustav von Pohl, von Helmholtz, and Count Reichenback. Perhaps their contributions to medical science will be better recognized in the decades to come.

Glossary

Throughout this book there are inevitably a number of acronyms which stand for frequently used nouns or important institutions. The list below amplifies these for the benefit of the reader, and also explains some of the technical words used.

Aetiology: the cause underlying a disease or disorder

ANSI: American National Standards Institute; the U.S. body responsible for physical standards in many areas of technology

ATP: Adenosine triphosphate; part of the essential oxygen transport system bringing oxygen to muscles. It leaves the oxygen there and returns as ADP, adenosine diphosphate, to repeat the trip

Attenuation: the progressive weakening of an electromagnetic field with distance from its source

Avogadro's number: the number of molecules needed to make up a gram molecule of any substance. It has the value of 6.02×10^{23} molecules. In this book it is used as a measure of ultimate dilution, so that beyond the Avogadro number there should theoretically be no molecules of a substance left in a solution which has been diluted by that number of times

BART: Bay Area Rapid Transit; the name given to San Francisco's electric train system

BEMS: the Bioelectromagnetics Society. Along with the *Journal of Bioelectricity* and its associated society, these organizations of scientists represent the foremost world research in this field

Carcinogen: any agent which causes malignancy in organic (living) creatures

Carrier wave: a wave which is used to carry a signal or information itself composed of very different frequencies. Imagine a matchstick being carried along by a seawave and deposited on the shore

CDC: Centers for Disease Control; the official U.S. monitoring body, based in Atlanta, which looks out for morbidity trends: they first spotted AIDS in 1979. *Published weekly*

CEGB: Central Electricity Generating Board; the U.K. body responsible for the 400 KV powerlines and generation of electricity. The Area Boards are responsible for 132 KV lines and below

CERL: Central Electricity Research Laboratories; the Leatherhead laboratories of the CEGB

CMR: cerebral morphogenetic radiation; this is my term for the electromagnetic signals which I believe the brain sends to the body in order to regulate its shape and to confirm the unique identity of the organism

Commissures: the name anatomists use for the connecting nerve fibres in the brain or body

Cosmic rays: these are not from the sun but from interstellar space and probably originate from other suns (stars)

CRT: cathode ray tube; a device which fires electrons at a glass screen which then lights up to show an image. Your TV has one

Degaussing: demagnetizing the metal hull of a ship or other metal structure

DNA: deoxyribonucleic acid, a long word, but it's quite easy to see how the name came about. An acid is any substance whose hydrogen potential (pH) is less than 7, Hydrogen potential being measured on a scale of 14. Acids have low hydrogen potential, because they tend to readily give up a hydrogen ion to nearby molecules and thus bind to them. In this way water-based chains, or polymers, are formed with any other substances which contain OH groups (water being H+OH, or H_2O, as we all know). A deoxyribonucleic acid is a chain of such acid molecules held together by a sugar (ribose, as in Ribena) and a phosphate; these long chains are like a biological electromagnetic tape, and contain information along their length which distinguishes them from any other creature's

DNA. This code is very long but completely defines the creature of whose shape it is one tiny component.

The structure of DNA is a twisted double helix (a very efficient aerial) whose two strands are held together by very delicate hydrogen bonds extremely sensitive to any form of radiation, which can change or mutate the DNA by fracturing the hydrogen bonds. Most textbooks neglect the electromagnetic nature of DNA

EEG: electroencephalogram; the minute changes in electric potential recorded from the living brain

ELF: extra low frequency; frequencies between 1-1000 Hz. Most domestic electric currents fall within this category, and are called power frequency currents

EMF: in this book it means electromagnetic field, but in ordinary electronics, emf often also stands for electromotive force

FDA: Food and Drug Administration; the U.S. body to which all food and health products must be submitted before being allowed on the market

Heaviside layer: over 60 kilometres above the earth's surface are distinctive layers which are capable of reflecting radio waves. The top two of these are called the Heaviside and Appleton layers after their discoverers

HSE: Health and Safety Executive; the body responsible for overseeing industrial and domestic safety in the United Kingdom. They do not consider that VDUs are damaging to health

Hz: Hertz; the name used, in honour of Heinrich Hertz, to describe the number of cycles per second at which any alternating electric or magnetic field alternates. So 50 Hz is fifty cycles per second

Inion: a specific part of the cranium

Ions: negatively or positively charged particles, which may have resulted from being knocked off their original atoms or molecules. Since nature always aims for equilibrium, they are therefore unstable, and will attach to any other particle of opposite polarity in order to achieve stability. Sometimes this is harmful to organic life, sometimes beneficial: negative ions are generally good for us, postitive ions bad for us

Laevorotatory: rotating clockwise as the field moves forward. The opposite of dextrorotatory. In stereochemistry two geometrically different molecules can be produced which are the mirror image of each other. One can distinguish between them by means of polarized light which is rotated to the right by one molecule and to the left by the other. All the main amino-acids only exist in the laevorotatory form, and are denoted by the prefix L-. For example L-dopamine

Mega: a prefix meaning a million. E.g., 1 MV is a million volts. A millionth of a volt by contrast would be a microvolt. Similarly, one micron (one μm) is a millionth of a metre or 10^{-6}m

Myelin: a fatty substance made of ingredients including melanin which acts as a sheathing insulation for nerve fibres. It also assists in conduction speeds

Milli: a prefix meaning a thousandth. E.g. a milliVolt is a thousandth of a volt. A thousand volts, by contrast, would be a kiloVolt

Morbific: disease-causing

Neoplastic: taking on a new shape. Cells which do this are not the correct shape, and therefore likely to become malignant or tumour-forming

NI-EMF: non-ionizing electromagnetic fields. Ionizing energy is so-called because it is powerful enough to knock electrons from the atoms which it encounters. X-rays are an example. Non-ionizing EMFs are not powerful enough for this, but may still cause biological damage in other ways

NRPB: National Radiological Protection Board. The body responsible for guidelines on EMF exposure in the United Kingdom

Ohm: a unit used to describe the resistance of any material to the passage of electrical current

OTA: Office of Technology Assessment; an office of the U.S. Congress which monitors and issues assessments of the general direction of technologies and their impact

Ontogeny: the process of coming into being

PEL: permitted exposure limit; the officially recognized maximum exposure levels at different frequencies. These vary from country to country

PMFs: pulsed magnetic fields; another way of saying alternating magnetic fields or transmissions which are emitted in short bursts

RAD: radiation absorbed dose; a measure of ionizing radiation

REM: Roentgen equivalent man; a unit of biological dose calculated to take into account both the absorbed dose (in rads) and the relative biological efficiency of the radiation

Refresh rate: the rate at which the screen of a VDU changes. If this is too slow it causes flicker. A normal rate would be 72 Hertz. but more screens now have higher refresh rates, say 90 Hz, which reduces eyestrain

RF: radio frequency; the frequencies used to transmit radio waves to your set, e.g. 500,000 Hz or 500 kiloHertz. An FM signal might be typically 100 megaHertz, which is a hundred million Hertz

ROW: right of way; the area nearest any power line where no one can legally build. The United Kingdom does not acknowledge the need for ROWs

Schumann resonances: the earth's own natural rhythms, which pulse at a frequency of about 8 Hertz, caused by resonances in the insulating cavity between the earth and the ionosphere. Schumann bands of ELF radiation spread between 1 and 30 Hertz

Side lobes: the stray radiation which accompanies a radar beam not aiming at the target direction

SIDS: Sudden Infant Death Syndrome; the layperson's term is 'cot death' though they are not strictly the same

VDU: visual display unit; a computer terminal embodying a CRT. The Americans call them VDTs, or video display terminals. Some terminals use non CRT displays, like gas plasma screens or liquid crystal displays (LCDs) which do not emit high fields

Volt: a unit used to describe the difference in electric potential between two materials

Watt: a unit used to describe the power of any electric current. Since electric currents are measured in Amperes, there is a fixed relationship between Volts, Amps, and Watts, so that Watts =

Volts × Amps. E.g. if a 6 Volt battery supplies a current of 2 Amps to a lamp, the power developed in the lamp is 12 Watts

1. In the Beginning was the Wave

If I start feeling there's anything dangerous I'll tell you. But it's not exactly danger I seem to feel about the place. It's–oh, I don't know–something oppressive like thunder: I can't tell what; but it worries me.

Richard Adams, *Watership Down*, 1972.

Yesterday, my newspaper reported (for the third time in as many weeks), that a rare migratory bird had lost its way and landed in Cornwall, and birdwatchers were gathering from miles around to glimpse it. The bird, a Blackpoll Warbler, usually flies two thousand miles south, guided by its sensitivity to the earth's magnetic fields, to overwinter in Africa. (1.1:68;153)

In Scandinavia unprecedented numbers of whales, which also navigate using the earth's geomagnetic fields, are beaching themselves along the coasts (1.2:12;37). Whole flocks of homing pigeons are reported to have lost their way back to their lofts. (1.3:254)

This morning I received a letter from a woman I have never met. She described how her husband, a sufferer from Alzheimer's disease, had suddenly and dramatically recovered his memory and clarity, just two hours after a brain scan which sent magnetic fields through his head.

A few weeks ago, from my electronic typewriter could be heard the conversation from someone's car telephone–to the great amusement of my son. Then a woman wrote to say she could hear men talking in her brain. She lives near Crystal Palace, where the BBC transmits its radio programmes.

Huge supplies of influenza vaccine were stockpiled last winter, because doctors know that influenza epidemics occur in cycles corresponding to the peaks of the eleven-year sunspot index, when electromagnetic radiation from the sun increases dramatically (1.4:133).

All these seemingly unrelated events indicate that the creatures of our planet, used to being affected by changes in natural geomagnetic fields, are also being affected by the artificial electromagnetic fields we ourselves have created.

All around us now–for the first time in human history–flows a mighty ocean of electromagnetic waves of myriad different frequencies and strengths. Night and day they are passing through the delicate cells of our bodies, which are simply not prepared by evolution to withstand their diverse influences.

We are not the only targets. The ozone layer, a mere one inch thick at sea-level densities, is being continually bombarded by the same invisible energies, from military and commercial radar, microwaves, radio waves, and the entire apparatus of the world's telecommunications industries. A single century ago none of this technology had been invented.

The Discovery of Electromagnetism

It all started just over a hundred years ago. In 1888, in what must surely count as one of the most important scientific discoveries ever, the young German physicist, Heinrich Rudolf Hertz, produced and detected the first artificial electromagnetic radiation. By sending a powerful electric charge across a gap and thereby producing a spark he was able to induce a smaller spark–evidence of the presence of an electric field–to jump across a second gap some distance away. This simple experiment has completely changed the way we all live; more than atom bomb, airplane, perhaps more than the invention of the wheel.

The existence of electromagnetic radiation had already been predicted in theory by James Clerk Maxwell, a Scot, in 1864. A few years later, in 1879, David Edward Hughes had noted that when an electric spark was produced anywhere in his house he heard a noise in his telephone receiver.

Hughes patiently traced the effect to the action of carbon granules in contact with a metal disc in his telephone transmitter. These granules were acting as a detector of the electromagnetic waves by sticking together slightly under their influence, so reducing the resistance of the mass, and producing a clicking in the receiver.

By 1882 Professor Dolbear at Tufts University had refined and publicly demonstrated the discovery. The use of electromagnetic energy never looked back from that date.

Radio Waves: the Race Begins

In 1885 the great Edison himself had successfully sent a 'wireless' message from a moving train, which had been equipped with a wire parallel to the telegraph wire strung along the track, by induction. In England W.M. Preece had performed a similar experiment. But Heinrich Hertz supplied the definitive proof. From that moment, as John O'Neill has pointed out, inventors like Tesla, Marconi, and Braun were quick to exploit this 'wireless telegraph' for commercial ends (1.5:194).

Hertz himself never lived to see the miracles derived from his discovery. He died in 1894 at the young age of thirty-seven, seven years before Marconi succeeded in sending a letter in Morse code between Cornwall and Newfoundland, thereby opening the first chapter in the amazing story of wireless telecommunications.

By 1912 slim volumes like Cassell's fully illustrated pocketbook *Wireless Telegraphy* was revealing to its readers the intricate secrets of Wimshurst machines, coherers, and the method of 'conducting light and heat by means of etheric waves'. Cassell's book quickly ran through three editions during 1913, and a further six during the Great War. Within its covers were advertised tappers, accumulators, and Leyden Jars 'for students and experimentalists alike'. In those days silicon and platinum detectors could be had for five shillings and six pence at distributors like Economic Electric Ltd of Fitzroy Square, London (1.6:41).

Electricity

The public interest in radio, the latest miracle from the new world of electricity, was intense. But before radio had begun to excite the imagination, electric power for domestic and industrial use had already changed the living environment so radically that peaceful scenes like Constable's *Haywain* (1805) were slowly fading from sight; the noise of industry had already taken command of the towns and could be heard distantly all over the British countryside.

To Yugoslav-born Nikola Tesla we owe many of today's electric advancements. One biography goes so far as to hail him as 'the main who invented the twentieth century'. What he failed to invent, he foresaw; from his fertile brain originated the incandescent light bulb, alternating current, the electric motors, which power so many of our modern domestic appliances, and the radio control of remote vehicles. Among his own claims he would add the wireless transmission of electric power, the EMP (electromagnetic pulse) death ray, and possibly even x-rays (before Roentgen in 1895).

In those early days few believed that anything other than direct application of electricity would be harmful to organic life. Although the application of an electric current, whether direct or alternating, had already been put to macabre use in the world's first criminal electrocution during 1890, and though Grover Cleveland was not allowed to operate electric switches in the White House for fear the United States might lose its President, such concerns were soon forgotten, and the headlong enjoyment of electricity and electromagnetism in all its diverse forms began in earnest during the first few years of the twentieth century. It was many years before the Curies discovered how deadly the field effects of ionizing radiation could be. In the case of non-ionizing electromagnetism it was to take decades, and be bitterly argued all the way.

Abrams and Boyd

Thanks to Lee de Forest's radio-wave detector, which replaced the crystal, radiotelephony became a reality in 1915, when spoken words were transmitted from Montauk Point, Long Island, to Wilmington, Delaware, some 214 miles away. Commercial radio broadcasting itself began five years later; KDKA in Pittsburgh began transmitting in November 1920, a year which had heard speech broadcast in Virginia intelligibly received as far away as Honolulu, nearly five thousand miles distant.

After this, thousands of amateur enthusiasts began assembling their own receivers at home. Albert Abrams, however, was assembling a very different kind of radio receiver. A well-heeled San Francisco physician, he was among the first to realize that the human frame could itself act as a receiver and possibly even a transmitter of radio waves. In fact, Tesla must also have known this, as years before he had observed that the high frequency electric 'streamers' he created with his device inevitably shunned the approach of any human being, suggesting that wave pressure of some sort was radiating from the human body (1.7:47). But it was Abrams who built an oscilloclast, a receiver tuned to specific frequencies to relate to specific human diseases. He discovered 'electronic reactions' in the stomach muscular tone of his patients, which changed noticeably when x-rays were turned on nearby, or when his instrument was primed with diseased tissue (1.8:3).

He was just in time. Dr Jean de Plessis, another early pioneer of research into the electromagnetism of living things, 'considered it doubtful whether Dr Abrams, if he had begun today [1925], could have discovered and developed his electronic reactions unless he first succeeded in screening his subject and apparatus from the various "interference waves" such as those emanating from radio stations'. How prophetic du Plessis' words were to prove!

As it was, Albert Abrams' brilliant discoveries were destined, like certain of Tesla's ideas, to be forgotten, despite the praises of none other than Sir James Barr, an ex-President of the revered British Medical Association (BMA).

In a remarkably forward-looking statement, which would have caused more than raised eyebrows if uttered even a decade ago, Sir James said:

When every important member of the community has a wireless telephone in his house and on his person, then medical editors and medical men will begin to perceive that there was more in Abrams' vibrations than was dreamt of in their philosophy (1.9:17).

Barr made this statement in 1922. It was not sufficient however, to prevent Abrams from being totally discredited, and he died, perhaps of shame, in 1924. Today, portable telephones are appearing everywhere, and Barr's prophesy is becoming true.

Even after Abrams' death the BMA set up a ghostbusting exercise. They planned a visit to the laboratory of one of Abrams' disciples, Doctor Boyd. Boyd's Emanometer, a similar device to that of Abrams, was to be tested by Sir Thomas Horder and a scientific team comprising Major H.P.T. Lefroy (Head of Wireless Research at the Air Ministry), Hart and Whately Smith from the War Office, and Dr Heald, medical adviser to the Director of Civil Aviation. An entrenched bunch, with a surprisingly military flavour, they fully expected to unmask the claims of the Glasgow doctor.

But Boyd was no charlatan. He had already foreseen the possible effects which stray radiation might have on his delicate instrument, and the laboratory was screened from floor to ceiling in copper sheet, just in case any radio signals crept in. In a series of twenty-five consecutive trials Boyd's blindfold operators distinguished correctly whether a specimen was present in the room or not. The chance of accidental success was one in 33,554,432, according to statistics. Despite this, since the inspecting party could not begin to explain the phenomenon, their report was savagely denunciatory (1.10:134).

Post-War Development of Microwave and Wireless Telecommunications

After the Second World War the microwave advances made during 1939–45 were redeployed for the peaceful purposes of commercial radio and television. Paul Brodeur, in his book *The Zapping of America*, succinctly describes this phase of microwave development in the United States.

A microwave radiotelephone system using line-of-sight relay towers was opened between Boston and New York in November 1947–the same year in which large-scale television broadcasting, also transmitted on microwave frequencies, got underway. By 1951, a coast-to-coast radio-telephone system had been established; it consisted of 107 hops, each about thirty miles long, and it used the tops of buildings, the peaks of mountains, and two-hundred-foot towers on the plains of the Mid-West. By 1960, more than a third of Bell's intercity telephone communication was being provided by microwave relay.

Brodeur chronicles the post-war impact of radio frequency energy on America:

Since the end of the war, the growth of sources generating microwaves and other radio frequency sources has been phenomenal. During the last thirty years [to 1977] the number of radio frequency transmitters authorized by the Federal Communications Commission — a figure that does not include military devices or transmitters — has risen from 50,000 to more than 7,000,000. The first microwave-telephone relay tower was built on Asnebumskit Mountain, near Worcester, Massachusetts, in 1946; today, nearly 250,000 microwave-telephone and television-signal relay towers — each with several microwave-generating sources — are strung across the United States...In 1945, there were only six television stations in the country, today, there are almost 1,000, all of them transmitting at either very high or ultra-high frequencies, and they are received by 121,000,000 television sets. At the end of the war, there were 930 radio stations; today, there are nearly 8,000 AM and FM stations. In addition there are about 15,000,000 Citizen's Band radio transmitters broadcasting on shortwave frequencies into homes and vehicles throughout the nation (1.11:35).

Brodeur's book, one of the most detailed accounts of micro-wave impact, was written in 1977, over a decade ago. Since then we have seen the massive rise of car telephones, early morning television, a steady rise in microwave ovens, VDU systems both in the office and at home, and the arrival of electronic mail, to mention but a few post-war additions to our electromagnetic world.

If the bombardment of our atmosphere with microwaves occurred with startling speed in America, the pace was hardly any slower in Britain and Europe. The first broadcasts from Alexandra Palace in London on 2 November 1936 were suspended during the war, but began again soon afterwards. In December 1949 Sutton Coldfield began BBC TV transmissions followed on 12 October 1951 by Holme Moss, and by the time Kirk o'Shotts and Wenvoe were in commission, in December 1952, over 80 per cent of Britain could receive monochrome television pictures.

During the succeeding decades transmission hours steadily increased as colour television and commercial television channels were added to the background level of radiation energy. More recently, satellite and cable television have increased this still further, in response to the public's seemingly insatiable appetite for electronic entertainment.

Few people have yet come alive to the chilling fact that most of the world's first AIDS victims were born in the same years as radio and television began. The details are dealt with in a later chapter, but in May 1988, after more and more disenchantment with the viral hypothesis of AIDS aetiology, two researchers discovered that UV radiation and sunlight could speed up HIV 1 proliferation (1.12:5;252). Only in October 1988 did it emerge that some HIV positive people never develop full-blown AIDS, nor is HIV virus found in all AIDS patients. (By

1989 *The Lancet* was steadily reporting sero-reversions–loss of HIV positive status).

1952 also marked the first sporadic outbreaks in North West London of the ailment myalgic encephalomyelitis (ME), often called 'Yuppie Flu', soon to be followed by an epidemic of this mild immune deficiency at the Middlesex Hospital in central London. Later in this book the connection between such disorders and electromagnetic fields is spelled out.

When Brodeur wrote his book less than 60 per cent of British households had a colour television. By 1988 90 per cent of British homes had colour television and 43 per cent also had a second monochrome set. What military installations were developed during the cold war of the fifties, accelerated by the nagging need to protect the West against sudden nuclear attack, is, for obvious security reasons, little chronicled. But as Brodeur says:

Besides the tremendous increase in the use of microwaves for radio and television broadcasting. . .hundreds of immensely powerful microwave transmitters–some with antennas that scatter microwaves from the upper layers of the troposphere, 7 to 10 miles above the earth–have been installed in the United States and overseas during the past twelve years [to 1977] to serve as links for civilian and military satellite-communications (1.13:35).

In 1976, moreover, at the other end of the non-ionizing electromagnetic (EM) spectrum, the 'Woodpecker' announced its presence. Robert Becker describes it:

During the US bicentennial celebrations of 4 July 1976, a new radio signal was heard throughout the world. It has remained on the air more or less continuously ever since. Varying up and down through the frequencies betweeen 3.26 and 17.54 megahertz, it is pulse-modulated at a rate of several times per second, so it sounds like a buzz-saw or woodpecker. It was soon traced to an enormous transmitter near Kiev in the Soviet Ukraine.

The signal is so strong it drowns out anything else on its wavelength. When it first appeared the UN Telecommunications Union protested because it interfered with several communications channels, including the emergency frequencies for aircraft on transoceanic flights. Now the Woodpecker leaves 'holes': it skips the crucial frequencies as it moves up and down the spectrum. The signal is maintained at enormous expense from a total of seven stations, the seven most powerful radio transmitters in the world.

Within a year or two after the woodpecker began tapping, there were persistent complaints of unaccountable symptoms from people in several cities of the United States and Canada....The sensations–pressure-pain in the head, anxiety, fatigue, insomnia, lack of co-ordination, and numbness, accompanied by a high-pitched ringing in the ears–were characteristic of strong radio frequency or microwave radiation (1.14:20).

Not long afterwards the United States began research into its own version of Woodpecker. The Project Sanguine would have made use of some two fifths of Wisconsin in the construction of a giant Extra Low Frequency (ELF) transmitter capable of being heard all around the world, particularly by submerged submarines. A committee which included Becker was set up to investigate possible biological effects and vetoed the concept. Nothing daunted, a new variant, Project Seafarer, was proposed. Again the system was halted. Finally, Project ELF was approved without Becker's evaluation, and now broadcasts at 16 Hz similar to the 10 Hz of Woodpecker, all around the world.

It is said that ELF transmitting systems are for military communications purposes, since they can transmit information to submarines even in the deepest parts of the ocean, which have hitherto been impenetrable except by whales and other sea mammals. In 1985 Becker wrote:

ELF electromagnetic fields vibrating at about 30 to 100 Hz, even if they are weaker than the earth's field, interfere with the cues that keep our biological cycles properly timed; chronic stress and impaired disease resistance result (1.15:20).

The frequencies that ELF systems use are very close to the natural brain rhythms of cerebrate creatures, the alpha and beta rhythms discovered in man by Hans Berger in 1929. The mass extinction of sea mammals such as dolphins and seals–even the curious new disorder in cattle known as 'spongebrain'–may not be entirely unconnected with this new kind of ELF transmitter, a variant of which, operating at 72 Hz, is right now being developed in Scotland. We, the public, simply do not know. All we know is that since 1976 the Woodpecker can be heard on any FM radio set. It is reliably said (by Becker) that the people of Russia cannot hear the signal since it is pointing only towards the West (1.16:19).

Electricity Consumption

Parallel with the rapid growth of microwaves and telecommunications has been the rise in electric power consumption, served by overhead and underground high voltage power lines. These lines also radiate weak electromagnetic fields. In 1920 only about 90 kW/hours of electricity per head of population in Britain was consumed. By 1987 the figure had grown nearly fifty times, each person now consumes a massive 4,400 kW/hours of electricity (1.17:86). This electric energy is distributed by 14,571 circuit kilometres of mains lines, of which nearly 10,000 kilometres are 400 kV power lines, not used before the early 1960s. In this way we are all bringing electromagnetic fields of myriad types into our homes, oblivious of the effects they may be having on our brains and bodies.

Apart from any effects of domestic electric appliances, which in Britain are mainly a post-war phenomenon, there are other important sources of electromagnetic radiation outside the home. Underground and overground electric railways for example. When scientists from Stanford Radioscience Laboratory were attempting to record geomagnetism at San Francisco in the seventies, they discovered that from 1972 the earth signals began to be disturbed by a mysterious interference. This was clearly man-made, since it didn't operate on Saturdays or Sundays, and for only twenty hours during the five weekdays. At first they thought the source might be the Stanford Linear Accelerator, but this proved not so. Then they moved their instruments 14 kilometres. Only then did they realize that the disturbance coincided with the new Bay Area Rapid Transit (BART) tube train arrivals and departures at Fremont terminal (1.18:129).

The BART system, like other rapid transit systems, works by passing current from a transformer-rectifier substation, driving the train's motors, and returning by the running rails. As the train moves, the size of this loop changes, as does the current, as the train accelerates it draws a large current, which lessens as the speed levels. When it decelerates, its dynamic braking returns power to the system, reversing the direction of the current. Together, this changing horizontal current loop generates a vertical Ultra Low Frequency (ULF) magnetic field concentrated at frequencies predominantly below 0.3 Hz. Because a heavily laden train can draw 7 MW at 1,000 volts DC, the current can be as large as 7,000 amps, leading to huge electromagnetic fields.

In an article in the *New Scientist*, Barry Fox quotes Anthony Fraser-Smith of Stanford University. 'The human body is an electrically conducting fluid–just a big sack of salty water. Any fluctuating magnetic field in a conducting fluid sets up electric currents.' Furthermore, cells have their own electric field, which would be affected by a varying electromagnetic field. There is a danger, argues Fraser-Smith,

that the large electromagnetic signals now being added to our environment may generate currents in the body which have long-term disruptive effects. No one monitors our total exposure to electromagnetic fields of all frequencies, and it is conceivable that the BART signals, although probably harmless by themselves, *may increase the possibility of harm from other electromagnetic signals* (my italics) (1.18a:89).

The influence on trees, certainly, is so strong that it can be measured simply by hammering two nails into the trunk a metre apart. A voltmeter connecting the two easily picks up BART signals, said Fraser-Smith. Just what the fields from electric trains may do to our tender brains is discussed in a later chapter.

There are other ways in which such fields are infiltrating our lives; some clerical workers in this age of desktop computing spend most of their day glued to visual display units (VDUs). As more firms switch to this sort of technology, millions of workers, mainly women of childbearing age, will be sitting in front of the screen, with an electron gun firing electrons at them all day. Highly-paid market dealers in the City of London and all other important financial centres face the same sort of screens. When one firm, Nokia Data, brought on to the market a VDU which it claimed emitted a much lower level of radiation, it quickly captured the lion's share of the market, illustrating the public concern over VDUs.

Ursula Huws, in her *VDU Handbook* (1.19:136), graphically illustrates the potential dangers; miscarriages, deformed children, premature and low-weight births, are reported from VDU operators all around the world. In Japan, the General Council of Trade Unions carried out a survey of 13,000 VDU workers of whom 4,500 were women. Of these 250 had become pregnant or given birth during the time they were working with

VDUs. Of these 91–more than one in three–had abnormal pregnancies, including eight miscarriages, eight premature births,
and five still births. What was interesting about the study was
the fact that the researchers studied the amount of time spent
at the screen by these women, and discovered a close correlation with pregnancy problems. Two thirds of the women who
spent more than six hours at the screen each day reported
problems, compared with just under half of those who spent
three to four hours a day at the terminal, and a quarter of those
who spent less than one hour a day there (1.20:138).

In many cases those same people work under fluorescent light,
another unsuspected electromagnetic hazard. Studies by Valerie
Beral from the London School of Hygiene and Tropical Medicine reveal a higher incidence of malignant melanoma among
workers exposed to this sort of lighting (1.21:23), results confirmed by at least another half dozen reports, according to a
review of fluorescent lighting hazards by the London Hazard
Centre in March 1987 (1.22:160).

Slowly the world's health authorities are becoming aware that
the benefits of electricity have a sinister counterpart; American safety standards for microwave radiation are being revised
downwards in a number of States. Unofficially, the New York
standard is only 50 μW/cm^2, even though the official American level (10 mW/cm^2) is a thousand times that of the Russian
maximum of 10 μW/cm^2.

Building near to power lines is also quietly being curtailed.
Studies are being carried out in European countries to examine
the effects of weak electromagnetic fields from all kinds of transmission lines, and electric machinery as diverse as fork-lift
trucks and electric blankets. The changes are being implemen-
ted with some incredulity and great reluctance. Meanwhile well-
documented cases of microwave irradiation, above average incidence of leukaemia in children living near power lines, occupational hazards for those working in electric-orientated jobs,
suicides, depression, meningitis and abnormal births continue
to surface.

The mysterious incidence of cot deaths, now the largest source
of infant mortality, and the suspicion that many immune-related
diseases, from ME to AIDS, are acquired from exposure to electromagnetic fields, is leading to a fundamental reappraisal of
the electromagnetic field intensities which at first seemed
innocuously weak. The story of the biological effects of electropollution is related in the pages which follow.

2. The Scientific Argument: Uncovering the Brain's Secrets

Advances in science are only possible with a broad outlook and a constant awareness of the possible relevance of concepts and techniques to branches of physics in which they did not first arise.

Dr Cyril Smith, *Electromagnetic Man*, 1989

The twenty-first century will soon be upon us. Anyone reviewing the twentieth must surely look back and marvel at the gigantic technological strides made during this hundred years. Humans have conquered interplanetary space; harnessed the atom for peace and war; can travel at speeds never before achieved on land, sea and air; and have learned to communicate instantly across the entire world. We have even managed to alter the very structure of life itself. What on earth, we might be forgiven for asking, is there left still to discover? The answer is very close to home: human beings themselves.

About 30 million Americans suffer from sleep-onset insomnia. Another 20 million are afflicted with high blood pressure. Untold thousands throb with migraines, wheeze with asthma, and scratch insanely at allergies of a myriad kinds. And now three million worried American citizens are carrying antibodies to one or more of the HIV viruses associated with AIDS, for which there is yet no cure. The number of AIDS cases doubles every fifteen months (2.1:9). An abysmal record for the most technologicaly advanced nation in the world!

The truth is, despite all this tremendous progress, we still do not understand how we come to be alive, or how we remain so. We do not even know how to control the shape of our bodies while we are living. Our brains, moreoever, with their ten-to-the-fourteenth-power cells, are an almost complete functional enigma to us.

Our bodies, despite the elimination of smallpox, tuberculosis, and poliomyelitis, are still ravaged by mysterious deadly
diseases over which we have absolutely no control and do not
understand. As a species we are too stressed, too aggressive,
and too diseased. Human life on our planet continues in consequence to be as nasty and brutish as three centuries ago, and
almost as short.

Early Brain Research

This book takes a completely new look at the way our brains
might work, and how they are trying to maintain our health
in the face of a rising menace–electromagnetism. For the brain
is the most exciting remaining frontier left open to human initia-
tive; its various anatomical components have been carefully
mapped for centuries, and every minute part of it has been dis-
sected, explored, irradiated, sectioned, and microscopically
examined.

Anyone exploring the shelves of medical libraries would be
amazed to see how many journals are devoted to brain research
and the neurosciences. More Nobel prizes have been handed
out for this branch of medicine than any other. But no one has
been able to explain how the brain works; we are like monkeys, playing in frightened yet curious amazement with a toy
clockwork mouse, when it comes to understanding brain
function.

The brain's role as a guiding principle was recognized by a
Greek, Alcmaeon, who lived in Crotona, that hotbed of intellec-
tual activity at the heel of Italy which gave Plato his best ideas.
But generally the Greeks regarded the brain as 'that thing in
the head' and left it at that, until Hippocrates began to chronicle its importance in his characteristic down-to-earth manner:

Some people think that the heart is the organ with which we
think. But this is not so. From the brain and the brain alone
come our pleasures, joys, laughter and jests...the brain is the
messenger of intelligence (2.2:128).

Hippocrates, though ignorant of the effects of airions, also recognized that the brain could be affected even by the changes in
the air, and in consequence that brain disorders were among
the most difficult for the physician to diagnose. How right he
was!

But his advice was soon forgotten. As Peter Nathan says:

At one time in Europe it was believed that the brain was nothing but a bag of mucus; and people thought that when one had a cold and one's nose became filled with mucus, a part of the brain was coming down the nose through little holes at the base of the skull (2.3:187).

A little before the great Plague and subsequent Fire of London, in 1664, Thomas Willis prepared us for a modern scientific study of the brain and its function by mapping its parts in detail in a book called *Cerebri Anatome* (2.4:226). Many of his quaint names for its components have survived. His illustrator was the young Christopher Wren, who was filling in idle moments before getting down to the (for him) more important task of designing the new St Paul's Cathedral.

Only about a century ago did Franz Josef Gall finally get down to the task of identifying function within these cerebral components. Since then we have made some progress, in the sense that we can now clearly understand connective relationships of some parts of the brain. But the greatest mystery of all life still stands before us. Sphinx-like; there is still no full explanation of brain function.

This may be because we have all been looking in the wrong direction. No one until now has put forward the notion that the brain is actually an organic, fully-operational radio transmission station, and not simply an electro-chemical transducer of wired information — a satellite telecom system, not an old fashioned telegraph.

The central tenet of this book, one which distinguishes it from other books on the brain, is the assertion that the brain is 'in radio contact' with every cell in its body. Indeed, electromagnetic influence is essentially what holds all body-cells in a stable condition–cut off your head, if you dare, and all your cells will start immediately to decompose into inanimate matter! Cut off an arm and, these days, you are likely to survive, and can have the arm re-attached.

Biologists studying marine creatures already know that something like this 'radio control' is happening.

Take the case of sponges for example. Certain sponges are colonies of separate creatures, and if you sieve two different classes of sea sponge into small pieces and leave them in a bowl

of seawater, a day later the two groups will have segregated themselves into their previous forms (2.5:131). The *Nanomia cara*
displays equally curious communicative skills. Like the Por-
tuguese Man of War, *Nanomia cara* comes from a class known
as siphonophores, because of the way they move through the
seawater they inhabit. *Nanomia cara* isn't one single animal,
since its cells are unconnected to each other; some cells push
it along, while others collect food for the group, and so on. But
the fascinating question, which points us towards a full explanation of the human brain, is how these creatures, way down
the evolutionary tree, communicate between their various cellular components so as to act in concert, even though not connected by nervous tissue. This fundamental example of radio
communication among separate cellular creatures is simply not
explained in the present literature.

If marine biologists had figured out the answer, we might have
begun to understand brain function much earlier. For the only
mechanism which we know could be available to siphonophores
to help them communicate is the phenomenon of electromagnetic field propagation and reception. And if creatures so primitive are using this method of keeping body and soul together,
isn't it possible that a complex specialized set of cells such as
those in a human body also do so, with one central radio station to keep the cellular population in touch with the latest
news? That central radio station is the human brain.

It seems that, way back in the course of evolution, single cells
were using electromagnetic fields to communicate. Indeed, glow
worms, conger eels, and certain other types of fish and mol-
luscs still do today, though of course they have now evolved
into multicellular organisms and, just like us; suffer serious
damage if their body's cells are amputated (2.6:12;225).

Pioneers

The idea of organic electromagnetic radiation is really not so
revolutionary a concept. We just haven't considered it in terms
of the brain before. Yet much of the supporting evidence for
such a view is already well known to physiologists and neu-
roscientists (2.7:142). It was discovered in 1929 that the brains
of all cerebrate creatures, including man, were continually sending out electromagnetic waves (2.7a:24). Amongst pioneers of
brain research Hans Berger stands out as the discoverer of these
brain rhythms (2.8:8).

He had been planning a career as an astronomer, until one

day he narrowly escaped death while on military service, when a gun carriage nearly rolled over him. That evening he received a telegram from his father hundreds of miles away, urgently inquiring after his health. Puzzled by the coincidence, he later asked why such an uncharacteristic telegram had been sent, for his father scarcely ever wrote to him: 'Your sister suddenly felt a remarkable concern that harm had befallen you,' was the reply.

Berger was so intrigued by this example of telepathy that he thereupon decided to make its explanation his life's work. To avoid ridicule, however, he conducted his investigations into brain waves in secret, and did not publish his findings until ten years after the actual discovery. He called the charts showing brain waves electrokephalograms (EEGs). The last book he wrote, *Psyche*, has never been translated into English but in it he speculated on the amount of energy the brain might need to transmit radio waves to other creatures (2.9:25).

Long before Berger's discovery, it was known that all living creatures generate small voltages which change dramatically when the tissue is injured or becomes active. But the significance of these, like brain rhythms, was completely baffling. This is not surprising, since radio waves themselves were only discovered forty two years before, in 1888.

Before then, in 1870, two Prussian Officers, Fritsch and Hitzig, had the brilliant but somewhat repugnant notion of testing the so-called Galvanic effect (the effect which causes frogs' legs to twitch when an electric current is applied) on the exposed brains of casualities of the Sudan war. They found that when certain areas at the side of the brain were stimulated by the current, movements took place in the limbs of the opposite side of the body (2.10:184).

Research information grew, and by 1913 Prawdwicz-Neminiski produced what he called the 'electrocerebrogram' of a dog, and was the first to try to classify his observations. At the same time Berger was discovering his 'electrokephalograms', in his garden shed, away from the eyes of his medical colleagues. Even when he finally plucked up courage to announce his results, few people were able to see the ultimate significance of these brain rhythms, and many doubted whether they were anything more than artefactual–produced by the instruments monitoring the brain–rather than by the brain itself.

Berger's work is summarized in Grey Walter's book *The Living Brain*, an account of brain research from its earliest days

(2.11:107). Even as early as 1929, Grey Walter's colleague at the Maudsley, Professor Golla, surmized that there would be variations in the rhythm of the brain in disease. This was subsequently found to be so, but even now it is not appreciated that the changes are telecommunicative.

The first convincing demonstration of the 'Berger rhythm' to an English audience took place at a meeting of the Physiological Society at Cambridge in 1934, using better equipment than Berger was able to devise.

Berger himself was rather unscientific and not a technical man, and knew nothing about mechanics or electricity. Furthermore, the equipment he was using, based on an out-of-date Edelman galvanometer, was crude. Yet it enabled him to detect differences in potential of one thousandth of a volt; and the line of his pen showed a wobble at about ten cycles per second (10 Hertz).

Adrian and Matthews developed superior apparatus which showed that the 10 c.p.s. rhythm arises in the visual association areas in the occipital region of the brain, and not in the whole brain (2.12:7). Before long it was realized that there may be tens of thousands of impulses woven together in such a complex manner that only the grosser movements are discernible with even our modern instruments. But in 1954 Grey Walter still had to admit that 'what makes these million cells act together–or indeed what causes a single cell to discharge–is not known. We are still a long way from understanding the basic mechanisms of the brain'.

Grey Walter himself was born in Kansas City in 1910, but came to England and went to Westminster School. At Cambridge he chose to study natural sciences, and continued with postgraduate research into conditioned reflexes, following the lines of Pavlov's research with animals in Russia. In his subsequent career Grey Walter charted many new kinds of brain rhythms, delta rhythms, alpha rhythms, and theta rhythms, as well as new facts about epilepsy. But the vital principle and purpose of the rhythms he charted so assiduously always eluded him.

In the same year as Crick and Watson were successfully investigating the structure of DNA (2.13:258), Grey Walter was writing *The Living Brain*, still in print today. Yet even in a later edition, published in 1961, he was forced to admit 'There are almost no firmly established facts about brain function; everything remains to be discovered, all problems are still to be defined, and every refinement of technique opens new possibilities of understanding and application.' But some facts were beginning to accumulate: 'We know that... studies of identical (monozygotic) twins have shown great similarities in their electric manifestations of brain activity.' This was early evidence of a direct link between the brain's signatory activity and the genetic structure of the organism to which it is attached.

Only the insight of how the brain monitors and repairs the genetic structure was lacking; regrettably neurophysiologists could only think in terms of line communications, when radio telecommunications were staring them in the face. If a light switch in the room is turned on, your brain's electroencephalogram (EEG) rhythms immediately change. Your brain has received and responded to a telecommunicating energy of which you may be completely unaware at a conscious level. Identical twins can influence each other's brain rhythms in the same way (2.14:80). Unfortunately few scientists have researched these phenomena, partly because, like Berger, they are afraid that their colleagues might classify them with the lunatic fringe, and they would suffer the ignominy which was an inheritance of Mesmer's 'animal magnetism'.

In 1982, in the absence of such research, Gerald Stern and Andrew Lees, consultant neurologists to University College London and the Whittington Hospital in London, writing about Parkinson's disease, could still only speculate:

It is quite possible that there are other critical substances and systems within our brains which to research workers are still as remote as the rings of Saturn. If each system can modulate the activity of the others by complex patterns of excitation and inhibition, it may be necessary to invoke the assistance of advanced computer technology before we can even begin to understand the normal control of movement, and its dissolution by disease (2.15:236).

I reluctantly leave aside the history of brain research at this point, for my purpose was simply to show just how much of an undiscovered country it is, and its importance as one of biology's last frontiers. The pioneers of the brain's territory–Karl Lashley, Wilder Penfield, Hughlings Jackson, Sir Charles Sherrington, Lord Adrian, Paul Broca, Brodmann, Karl Pribram, Hans Berger, Grey Walter, Edward Hitzig, Pavlov, Galvani, and others, deserve a complete book to themselves.

Brain Function and Electromagnetic Energy

We can now begin to understand what electropollution does to us all, by starting with some electronic physics. Don't worry! It won't be too painful. In fact it is so obvious that its principles will already be familiar to many readers.

Physicists know that if two metallic plates, adjacent but not touching, are each connected to the positive and negative terminals of any ordinary battery, positive and negative charges will build up on the surface of the plates. The nearer these two plates are physically to each other, the stronger the charges, or potential difference, between them will be. So if you then move the plates to and fro, the charges will increase and decrease, causing electrons to flow down the wires to which they are connected, first one way then the other.

If these electrons are made to change direction fast enough, the associated magnetic and electric fields spinning around the wire do not seem able to keep up with the change, and get trapped outside the wire whose electrons have, as it were, locked them out by their change in direction, before they have time to collapse back into the wire. Left outside with nowhere to go, the flux of electromagnetic energy does a very strange thing; it moves away from the wire at the speed of light, gradually weakening as it goes. Physicists call this process attenuation. The electric field travels in a plane parallel to the wire, and the associated magnetic field travels in a plane perpendicular to the wire (2.16:63).

This is the answer to Berger's question of where does the brain's energy go; it goes to electromagnetic transmitters in the brain. Modern radio transmitters today do the same when they are sending radio and television signals to your set.

It seems that the brain, which oscillates constantly during its life, is also acting as a radio wave propagating transmitter. In this case the 'wire' is formed by the million or so nerve fibres connecting the two halves of the brain (the cerebral hemispheres). In each hemisphere are millions of tiny plates called great pyramidal cells or Betz cells, after Vladimir Betz, who discovered them in 1874. The Betz cells are connected by a one-to-one fibre commissure, each insulated from its neighbour by a myelin sheath. These commissural or connecting filaments are noticeably harder than other brain tissues, which is why Thomas Willis called them *corpus callosum*, 'hard body'.

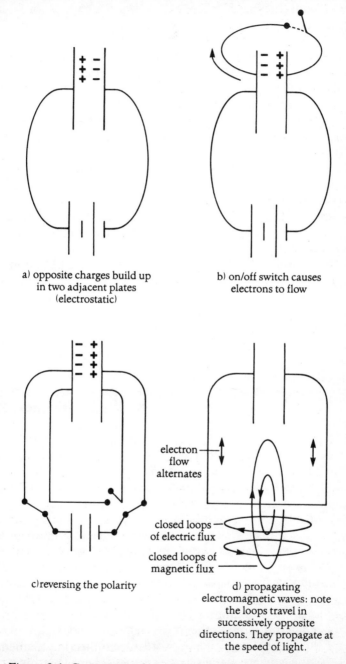

a) opposite charges build up
in two adjacent plates
(electrostatic)

b) on/off switch causes
electrons to flow

c) reversing the polarity

electron
flow
alternates

closed loops
of electric flux

closed loops of
magnetic flux

d) propagating
electromagnetic waves: note
the loops travel in
successively opposite
directions. They propagate at
the speed of light.

Figure 2.1: Converting electrostatic energy to electromagnetic
energy in the brain.

Such a mechanism implies that the brain is capable of sending out simultaneously as many signals as there are pairs of
Betz cells in the cortex, a radio transmitter of formidable complexity. What we measure of this process with our modern EEG
instruments is a crude and insignificant reflection of the actual
signals being transmitted; it's as if the brain had a million or
so radio stations in operation simultaneously. Scientists will
surely need a spectrum analyser and computing system of
phenomenal power to disentangle such signals meaningfully.

This, in brief, is the ultimate function of all cerebrate creatures' brains, and the evolved equivalent of the intercellular communication system used by *Nanomia cara*, and the red and
yellow sponges. Other variants of the mechanism explain how
fireflies also give off electromagnetic signals, in their case visible, because they happen to use the frequencies of visible light.

It is probable that many other insects use similar communications techniques. Bees, for instance, can distinguish anomalies in the earth's weak magnetic fields. I would hazard a guess
that insects' and butterflies' antennae are not only for smell-
ing (the currently accepted view) (2.17:256). The received opinion that insects can smell a pheremone from the female several
miles away seems suspect, in that the dilution of her scent
molecules in air so distant must exceed Avogadro's number in
such cases. This means that technically there is not a single
scent molecule left in the air so far away from the source. Philip
Callahan pointed to a possible infrared sensing mechanism in
insects as early as 1965 (2.18:38).

It is more likely that an electromagnetic signalling system is
at work, which signals her presence to the male. In fact French
scientist Fabre, in his book *Moeurs des Insectes*, relates how male
Great Peacock butterflies were still able to track down the female
even after Fabre had covered her with a variety of diffused smells
to put them off the scent. Interestingly, when he put her inside
a bell glass the males completely ignored her, even though she
was in full view. Glass does not allow the passage of UV radiation, which suggests that the radiation from the female was in
that frequency range, just above the visible spectrum (2.19:151).

Ants can sense a blockage in their marching column well
before it can be communicated feeler-to-feeler, since the sold-
ier ants come running up the line to remove obstructions well
before they have been told by word of mouth (so to speak)
(2.20:166). There are many other examples of electromagnetic
communication, not only between the cells of a single organ-

ism, but between separate organisms (2.21:37;244) (2.22:36;245) (2.23:173) (2.24:267). It is these mechanisms which the use of our newly discovered electromagnetic energy may be disrupting, with serious consequences, as we shall see in subsequent chapters. Last year many thousands of homing pigeons were lost one weekend when a huge solar flare occurred (2.25:245) (2.26:183), though some commentators blamed a military microwave system (2.27:255).

In the early days of radio, engineers found they could influence an aerial's direction by adding waveguides. As D.C. Green writes:

> The directional characteristic, or directivity, of a transmitting aerial is very useful because it allows most of the transmitted power to be sent in the wanted direction, and very little in other, unwanted directions. This reduces the transmitter power required to produce a given field strength at a given point in the wanted direction (2.28:106).

Nature doesn't miss a trick, and sure enough the brain has a waveguide. The lateral ventricles immediately below the corpus callosum ensure that most of the radiated signal is given downward directivity, and very little goes upwards towards the skull. This front-to-back ratio, as it is called, means that the propagating system can operate efficiently at lower strengths. When one

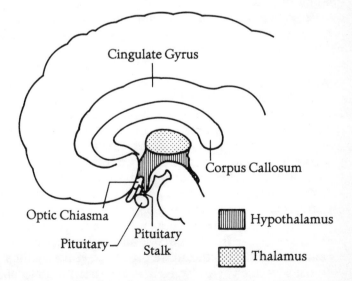

Figure 2.2: Cross-section of brain showing hypothalamus and thalamus.

considers that the brain consumes as much as 40 per cent of
the oxygen we breathe, and is the first port of call after the lungs,
(ensuring that it gets first choice of freshest air), the efficiency
is underlined.

The 'gain' of any aerial is a measure of its directional properties, and indicates the extent to which the aerial receives signals better from some directions than others. The gentle curve
of the corpus callosum improves its gain enormously. If it were
flat the system would not be quite so efficient. No one has ever
explained its shape in this way before, but this anatomical feature certainly helps to support the notion of the brain as a radio
transmission (and reception) system.

The thalamus, which is clearly a delicate switching device,
controlling the individual polarity of the large pyramidal cells
on the surface of the cortex (2.29:28), lies immediately below
the lateral ventricles and the corpus callosic sheet. In that position, this switching device can simultaneously control and
monitor the effects of the propagated waves. If it were anywhere
else in the brain, above the corpus callosum, say, it would not
be able to carry out either function so well.

The fibres coming from the thalamic nuclei to each large
pyramidal or Betz cell enable it to control the polarity of a single cell. It is also known to exercise a general control over the
amount of sensory and motor activity going on in the body.
It does so by means of its assistant, the hypothalamus, which
sits a little way out of the firing line. One sign to the hypothalamus from the boss and the instruction is relayed in turn to the
pituitary gland, and the motor and sensory systems start to shut
down (2.30:56).

The reason for shutting down in this way from time to time
is obvious. We lose some 500 million cells each day, and they
must be replaced if we are to continue to function. By shutting
down all the electromagnetic 'noise' emanating from muscles,
movement, sensory stimuli and so on, the vital repair signals
from the brain can be transmitted more clearly. Some life-
support systems must be kept going, and this is the job of the
pituitary gland, discussed in Chapter 3.

What can the brain, which contains not a single muscle, need
all that oxygen for, if not for developing the energy to transmit
its radiations? Indeed, the air we breathe goes first to the brain
before it goes anywhere else, so that the brain has first call and
the freshest of all air and oxygen.

The briefest cessation of oxygen (anoxia) can cause irreparable damage to the brain, affecting the mental ability of infants, impairing judgement, perception, and psychomotor performance in adults, and with noticeable damage to brain cells. By curtailing the amount of oxygen to the brain–cigarette smoking and living in inner cities both have this effect–we could be stunting our growth and reducing the signal strength to our bodies.

The physical structure of the corpus callosum also accords with its role as an aerial; it is noticeably harder than and different from the rest of the brain's tissues. One would expect that any filament which is subjected to constant alternating currents, an electric fire's filament for example, would need to be hardened against such use. Each filament in the callosic sheet is also covered with a myelin sheath, so that it is effectively insulated from its neighbours. This would be a necessary adjunct if clarity of transmission was of paramount importance for each individual filament. (There even appear to be mechanisms in the brain to dampen down the strength of unwanted cortical signals, called neurosuppressors.)

Such a view of how the brain works would predict that the callosic sheet is prone to become somewhat hotter at times when signal strength is at its most powerful; when the body is in need of repair or is under attack from viral infection for example. The most common symptoms of such attacks are of course temperature, headaches and fever, which accord with the expected result.

Another fact which has emerged from brain research is that the brain wave patterns of two different individuals are never alike, just as their DNA macromolecules are also unique; the trained person comparing many EEG records can often match up traces belonging to the same person. This is what one would expect if the purpose of the rhythms was morphogeneic, that is, to control the shape of our bodies, no two people, even identical twins, are exactly alike, as we know from studies of fingerprints (2.31:48).

So far the argument is fairly well supported both by physics and the anatomical lay-out of the brain. But the concept is still only an outline, and will only gather convincing strength from a good deal of further refinement, supporting evidence, and most important of all, its ability to make correct predictions.

A first prediction is that the most important signals of all come from further down the central longitudinal fissure, immediately above the cingulate gyrus. The reason for suggesting this arises from the way the cortex itself is laid out. Back in the thirties Karl Lashley, Wilder Penfield and their colleagues mapped out the sensory and motor areas of the brain (2.32:199), and were suprised to find that these were arrayed spatially on the cortical surface, with the parts of the body furthest away from the brain represented nearer and nearer to the central fissure which divides the two hemispheres. This was because the signals which had to travel furthest, to the feet for example, needed plates on each side of the cortex nearer than those for signals which had to travel a shorter distance.

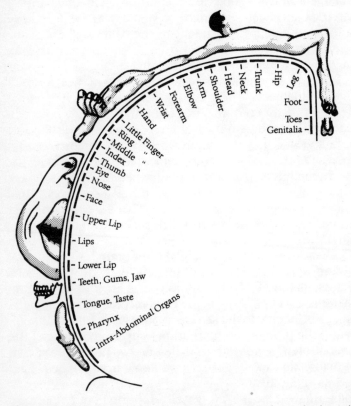

Figure 2.3: Penfield's homunculus. He mapped out the brain's localization of function, but did not realize it was arrayed for maximum propagation.

The signals relating to the sex organs though, Penfield discovered, were displaced and instead of being about half way across the cortical surface were a little way down the actual central longitudinal fissure itself.

Given the relative importance of the procreative drive, this exception seems very understandable. Even more important, presumably, is the basic signal which distinguishes 'us' from 'not us', that is the immune defence signal. But Penfield was scarcely aware of immunology in the thirties, and did not find it easy to explore even further down the fissure.

I suspect that from deep within the central fissure–the closest plates of all and therefore the strongest radiating components– there come the most powerful and most basic signals of all, those which make the primary distinction between 'friend' and 'foe' cells. This is, I believe, the seat of the cortical control of the immune defence system, a system mediated not only through hormonal action but also by electromagnetic means. If my hypothesis is correct, any damage to the areas close to these cingulate gyrus cells, buried deep as possible in the recesses of the cortex, will result in immunodeficiency of one kind or another.

Neurophysiologists tend to keep quiet about the cingulate gyrus, because they have no explanation for its function, apart from saying that it is associated with the hypothalamus, and forms part of the limbic system, and was therefore among the earliest part of the cerebral hemispheres to develop. The other parts of the limbic system are the hippocampal gyrus, the uncus, and the amygdala (from Latin words for 'sea horse', 'hook', and 'almonds' respectively).

The hippocampus and its commissures seem in some ways like a minor version of the hemispheres and are situated on either side of the brain in the temporal lobes. Its proven association with memory makes one wonder whether it is part of the receptive system, receiving signals back from individual cells for integration with the brain's transmission system (2.33:53). We know so little about this set of components that it is not easy to say. But as for the cingulate gyrus, most textbooks are silent. Though it appears quite different from the cortex, neither Berger nor Grey Walter gave it a single mention; Morgan and Stellar honoured it with one sentence, and Peter Nathan, writing his second edition in 1982, does little more.

All the evidence from lower mammals and reptiles suggests that the cingulate gyrus is older than the cerebral cortex, but still part of the new brain. It seems to be associated with emotion, but its function is still mysterious. According to CMR theory however, its anatomical position suggests that its signals are of the utmost power and importance, and may be connected with the signalling mechanism I have outlined.

Space prevents me from detailing the mass of further evidence which supports this way of looking at the brain. Others have offered similar observations during this century, for example George Washington Crile, (2.34:64), Guyon Richards (2.35:214), and George de la Warr, (2.36:257). I now turn to the way the cerebrate creature protects its radio signals from outside interference.

Outside Interference
The Skin
Until the beginning of this century for the entire previous history of the planet the only competition for organic signals were the sun's daily radiation, and the earth's geomagnetic field. Whether the moon and planets also have an influence is outside the scope of the book, but should not be discounted. After all, the moon effects the ion content of the earth's atmosphere each month, and when it is full presses its ionic influence closer than usual upon the earth as a result of the sun's gravitational pull.

Clearly such external radiative influences might easily 'jam' the system. Accordingly, some cells have been specialized into acting as shielding protectors, forming the dermis and the epidermis. Normally this protection is sufficient to ward off ultraviolet radiation from the sun's rays. Occasionally however psoriatic conditions appear; this can temporarily be cleared by applying UV radiation at specific frequencies, and infantile jaundice is another condition cured by the emerging science of photomedicine.

All these problems are evidence of an electromagnetic aetiology of cellular disorders, of which this book will provide many other examples. Skin cancers in Australia are now becoming a serious health hazard as the synergistic effects of natural and artificial electromagnetic energy combine to permeate dermal cells not evolutionarily prepared for such insults.

In short, our skin is designed to protect the organism not only from penetrative attack, but also from ultraviolet radiation, in the latter case containing melanin to ward off the sun's rays. Without such protection ultraviolet or other radiations would mutate the interior cells of our bodies. It is known that UV irradiation kills bacteria and causes the hydrogen bonds in DNA to fracture (2.37:217a).

Interestingly if someone unfortunately amputates a part of their body, thus exposing its interior to external radiation, the amputated part can sometimes be regenerated. This can only happen if the skin is not sutured, that is not sewn back together to cover the exposed area of amputation. Furthermore, the speed of regeneration is improved by the weak local application of electromagnetic, or even simply magnetic fields. This technique was known as early as the tenth century A.D. and is mentioned by an Arabian physician, Ali Abbas, in his *Perfect Book of the Art of Medicine*. So our skin seems to have a role of protecting the internal cells from radiative influences.

Melanin and its associated melatonin are present when external radiation needs to be warded off, which is why we get sunburn or have dark skin. Perhaps the melatonin-serotonin cycle works in the same way internally to suppress or enhance at certain times any radiative influences on the body's internal cells. Since the cycle is triggered by the presence (or absence) of light this would make sense. Melanin is also an important ingredient of the myelin sheathing which surrounds our nerve fibres, and acts not only as an insulator, but also helps the speed of nervous conduction (2.38:21).

Multiple Sclerosis

Multiple sclerosis is a condition where this myelin is progressively stripped by an unknown mystery agent from the central nervous system (2.39:82) (2.40:11). Jane Clarke was probably right to suggest that the mystery agent is none other than electromagnetic energy (2.41:50). As the myelin disappears it invokes a defence mechanism whereby the immune system seems to regard the inevitable radiation from the body's own conducting nerve fibres as 'foreign', and creates defensive plaques against it.

As Barry Bloom wrote in *Nature*:

The histopathologic picture of the demyelinated lesions closely resembles that of cell mediated immune reactions, with lymphocytes and macrophages predominating. However, the often acute onset of symptoms and rapid clinical remissions suggests that mechanisms which are not characteristic of cell mediated lesions. . .are contributing to the disease (2.42:33).

One of the worst cases of multiple sclerosis I have ever observed was a young executive living inside a triangle bounded by three electric railways. Another curious fact of this disorder is that it rarely occurs in countries like Africa, where the indigenous population is acclimatized to solar radiation, again implying an electromagnetic aetiology. If however the population is suddenly irradiated, even at very low levels, when they are not so acclimatized–as in Iceland and the Faroes when radar was installed–outbreaks of multiple sclerosis occur.

These geographical clusters have at least forced medical science to eschew offering a strictly viral aetiology. Bryan Matthews in his excellent review of the disorder says:

I must make it immediately plain that there is no evidence at all that MS is catching, or that it can be passed on by any form of contact.... If MS was transmitted in this way an increased incidence in the spouses of those with the disease would be certainly expected, but it is no higher than chance (2.43:169).

I have not space here to deal with the matter in the depth it deserves, but I note in passing that the cerebral damage seen in multiple sclerosis also often involves the corpus callosum. Moreoover, like cot death, its incidence and importance has grown with increasing use of electricity, from its first discovery in the nineteenth century by Charcot (2.44:45), the incidence of MS has now grown to some 50,000 in the United Kingdom in any one year–many times the incidence prior to World War Two.

Here perhaps is yet another example of the importance that our bodies attach to maintaining signal integrity. The immune system can even distinguish between its own unique signal and the signals coming from its own neural impulses. Using serotonin to enhance and melatonin to suppress it can optimize the signals while CMR transmissions are in progress. When the myelin is stripped from the nerves the melatonin is no longer able to exert its insulating and shielding influence (2.45:72), and the nerves, treated as a 'foreign' radiation, just like any bacteria, are exposed to the onslaught of the body's immune defences. At present medical science has no explanation of multiple sclerosis, but by following the line of reasoning I have offered, many of its bizarre symptoms become understandable. Removing MS sufferers from any sort of electromagnetic energy should lead to remission.

DNA

Finally, the physical structure of DNA in each cell is curiously like that of a radio set's internal aerial, as if it were indeed designed for that purpose. It is in the form of a tight coil, with a complete turn every 3.4 nm and with ten chemical bases (of four types) in each turn. These DNA bases are called adenine, guanine, cytosine, and thymine, and the two stranded 'backbones' of its double helical structure are made of sugar phosphates which thereby hold them together (2.46:10).

Why should they be twisted in that way? No one has yet asked let alone answered this question. But the essence of an electromagnetic field is that it is what is called laevorotatory, and light is thus often regarded as circularly polarized. With a circularly polarized field the electric and magnetic vectors generate complementary chains of a double helix, just like that of DNA. In other words, DNA is eminently suitable to receive, indeed to transceive, because any aerial is also a transmitter of suitably modified electromagnetic waves. So at the heart of every cell within our body lies a little radio receiver, tuned by its unique sequence of chemical bases to receive one and only one unique electromagnetic signal.

This mechanism comprises the other part of the system by which morphogenetic radiation is propagated and received. There are grounds for suggesting that in fact one of the DNA strands might be acting as a receiver and the other, whose bases are complementary, as a transmitter receiver. To quote D.C. Green:

The performance of an aerial is the same whether it is used for transmission or reception, the main difference lying in the magnitudes of the powers involved. The power handled by a transmitting aerial may be very large, perhaps several kilowatts, but the power absorbed by a receiving aerial is very small, possibly only a microwatt or two (2.47:106).

Of course, he was writing about artificial radio transmitters and receiving systems, and the power values he mentions bear no relation to organic systems.

If cells are communicating by means of electromagnetic radiation, we would expect to be able to measure the radiation as it comes from individual cells. This has in fact been done. The energy for such radiations appears to be supplied by the mitochondria, which is like a battery found within each cell but situated outside the nucleus itself. If power supply to the mitochondria is severed the radiative strength of the cell gradually dies away. We can measure this waning radiance by means of Kirlian photography, a technique whereby the cell's own energy is mapped by high frequency electromagnetic pulses which even show parts of the organism which are not present (2.48:193). This initially unexpected result is not so surprising in the context of morphogenetic radiation; if each cell carries a blueprint of its entire form, then Kirlian photography may have the ability to recreate that entire form from only a portion of the organism.

Eventually this may be done as a routine investigation by police forces, who are already using DNA 'fingerprinting' to identify individuals suspected of crime, since technically the DNA contains the physical blueprint of the organism whence it came, and can presumably be reconstituted.

Before the form of any organism can be recreated fully in this way we must learn to interpret the signal codes which emanate from the brain, and are duplicated at micro-level by each cell. It all sounds far-fetched, but in fact is little different from the way we can already recreate a film by means of a video cassette.

We can at least now begin to see a mechanism whereby electromagnetic waves are being radiated from the brain, which can inhibit or encourage DNA in its synthesis of protein, or can reinforce the correctness of its unique 'callsign'. There are also grounds for believing that DNA itself contains a mechanism for locally re-broadcasting the callsign it contains; and finally that this callsign is comprised of a digital equivalent of the whole organism which it represents. We are therefore beginning to describe a system for the control of any cerebrate creature's shape.

Pausing for one moment to consider this, it seems a natural progression of evolution that such a system should exist. Way back in the primeval depths of time cells may have banded together for self protection. They would need a communications system to do so which distinguished them from alien influences. Presumably at the outset all primeval cells within one organism were, like sponges, undifferentiated. With evolution cells must have started to take on specific roles, and at that point a central controlling mechanism became necessary, to maintain order and see that the specialist cells knew what was required of them. Also from that point it must have been necessary to distinguish members of the 'club' from uninvited guests. Thus the need for an immunological system arose.

The 'callsign' system, which I call cerebral morphogenetic radiation (CMR) then came into existence. Possibly for this reason its control is buried deep in the central longitudinal fissure, below the falx cerebri, the gap which separates the two hemispheres in our modern brains. At any rate this unique callsign serves a dual purpose, morphological and immunological. We started by looking at cerebrates as electromagnetic rather than electrochemical systems. There is, of course, no need to regard the two systems as mutually exclusive.

Plants and Radiative Cellular Emissions

What of plants though? They do not have brains. Nor do they have muscles. Consequently they have no need to carry oxygen to the muscles in the way that animals do. Animals carry oxygen to muscles by making use of adenosine diphosphate (ADP) and adenosine triphosphate (ATP). The ADP forms a weak bond with an oxygen atom, thus becoming ATP, and the oxygen is then transported to the muscle and deposited there. ATP itself is formed from glucose, but if dropped on an isolated muscle, it will make it contract, which glucose alone will not do (2.49:10). The ADP-ATP transport system works very well. This system too might be electromagnetically controlled, but not by DNA, for a very special reason. If DNA were to make use of a signal containing Adenine (or its complementary base Uracil, which is found only in RNA), it might confuse and swamp the oxygen transport system.

DNA does not ever contain Uracil, and for this reason we can never manufacture certain 'essential' amino acids which we need to survive. These essential amino acids, eight of them, have to be ingested from plants. Could it be that the Uracil 'signal' — which is never found in DNA, only in RNA — is a forbidden signal, because it would interfere with the oxygen-carrying operation?

Plants, unlike animals, do not defend themselves against radiation from the sun; they make use of it via photosynthesis, to develop their structure. That plants also emit radiations, nevertheless, has also been discovered for some decades. Gurwitsch, a Russian biologist working in the thirties, called this 'mitogenic' radiation. He detected it in the growing tips of onions, and it seemed to initiate cell multiplication and division (2.50:112). Gurwitsch's experiment supports my hypothesis, in that it demonstrates a primitive form of the sophisticated system I have postulated for cerebrate creatures.

Other scientists have replicated his findings. In 1930, Cremonese made direct photographic records of radiations emitted from human subjects, as well as from their saliva or blood specimens. This ultra-weak biophoton emission from living cells has been extensively studied by Franz Albert Popp, who found that cells give out a high burst of such energy just before they die (2.51:208), and by Dr Eugene Celan.

In 1986 at his Bucharest laboratory Celan took a culture of tumour cells growing in darkness under a quartz cover slip and put on its top a drop containing actively dividing yeast cells. He was able to show that radiation from the yeast cells passed through the quartz and killed the tumour cells (2.52:42). If he used glass as a separator the effect did not occur. Since glass will not allow ultraviolet radiation to pass through it, the probability is that these radiations are in the UV range, like the frequencies found by Fabre.

Similarly, in 1976, Kaznacheev and Shurin of the State Medical School, Novosibirsk, Russia published the results of their work on distant intercellular interaction, for which they received one of the USSR's highest scientific awards. They too were able to kill one organism by radiative means from another to which it was toxic, with the two separated by a quartz barrier (2.53:143).

There can be little doubt nowadays that cells do emit coherent radiated information to each other, not only for communication, but also for aggressive objectives. Georges Lakhovsky, an early advocate of the radiative approach to biology, called disease 'a war of radiations', and Albert Abrams had previously shown that the waveform of malaria could be inhibited completely by the waveform of quinine (2.54:3). The pattern of one on his oscilloscope was the mirror image of the other, a most significant pointer for the medicine of the future. As we will see later, it is not for nothing that most of the ailments which afflict AIDS patients and others suffering from immune deficiencies of various kinds are caused by opportunistic monocellular invading organisms — which have no need of intercommunication — rather than multicellular organisms.

I cannot do justice to all the evidence available to me for promoting the notion of cerebral morphogenetic radiation in this small book. I have sketched the outlines as a preliminary to the main purpose of understanding and minimizing the adverse effects of electropollution. However, one other curious unsolved biological mystery deserves mention: the fact that illness of psychological origin often has physical results. It is often said, for example, that 80 per cent of cancer patients suffered a serious stress trauma within two years previous to the arrival of cancer (2.55:232).

Psychoneuroimmunology

In the last decade the concept of psychoneuroimmunology has made great strides. It has done so, however, always within an electrochemical framework. Psychoneuroimmunology (PNI) tries to answer the fundamental scientific question of how the mind can manipulate, at a distance and outside the blood-brain barrier, the intricate biochemistry that determines the course of disease. Whereas holistic therapists talk about the mind healing the body, PNI researchers investigate how the brain affects the body's immune cells.

Their answers, however, are all couched in the conventions of chemistry rather than electronics. They are the chicken, I believe, to my egg. Chemical substances are second to the electronic forces which bind them together, and chemical and electronic systems are not mutually exclusive. Though, like me, PNI researchers believe that the brain and the body are a closed system, any explanation in chemical terms is immediately confronted with certain difficulties; action at a distance is not possible with a chemical system. Yet there is clear evidence of action at a distance in the emergence of cellular or cerebral dysfunction and disease. Thus, though PNI research quite rightly identifies that the immune system is ultimately under cortical control, it proposes that the mechanism is controlled by hormones issued from the brain, and more recently — since the system is obviously two-way — by hormones issued from the cells.

Gerard Renoux from Tours University, France, found that destroying part of a mouse's cortex (which does not significantly affect the animal's behaviour) changed the structure and activity of the immune cells. Besedovsky discovered that by implanting electrodes in the brain of a rat, and then injecting the animal with foreign cells which stirred its immune system into action, the electrical activity in the rat's brain increased, and the level of some important brain chemicals dropped temporarily (2.56:262).

But how do the lymphocytes know where to find the offending bacteria or viruses? Could it be by a kind of electromagnetic sensing mechanism? If lymphocytes are indeed programmed specifically by electromagnetic signalling at a distance, it would make sense of a large part of immunology which currently is obscure and incomprehensible. Ed Blalock of Alabama University believes that hormones provide the essential communications medium between body and brain. In order to do so he has had to show that interferons can actually create the hormones previously manufactured exclusively by the brain. Other research, however, shows that injection of interferons causes EEG changes before chemical ones. And work by two English scientists, Smith and Aarholt, shows that the production of endorphins can be initiated by electromagnetic radiation at specific frequencies at a distance (2.57:230).

Interest in psychoimmunology has increased so much as a result of the AIDS pandemic that in 1983 the National Institutes of Health budgeted US $11 million, most of it for studying the behavioural aspects of AIDS. The medical world is therefore turning its attention to the relationship between the mind (or rather the brain) and the cell. Such substances as neuropeptides, which seem to regulate the immune function, yet are produced by emotion in certain cases, are claimed to play an important part in AIDS.

Certainly when AIDS victims are told they are antibody-positive the depressive effect will contribute, like all stress, to a lowered immune defence mechanism. But this does not explain why, in AIDS, changes in the brain's rhythms begin even before there are any overt symptoms (2.58:130) (2.59:60). By contrast it is relatively easy to show that electromagnetic radiation, particularly CMR, can induce subsequent and quite specific increases in protein synthesis (2.60:79). I discuss this later in more depth when reviewing the role of sleep.

In suggesting this radical departure from the conventional biological paradigm I am reminded of Thomas Kuhn's words that those who change the paradigms of science are 'either very young or very new to the field whose paradigm they change' (2.61:150). Kuhn suggests that following a period of 'normal science'–such as the development of mechanics, for example, after Newton and his *Principia Mathematica*–scientists turn from puzzle-solving to worried discussion of fundamentals, because some crisis has arisen which they cannot accommodate within accepted theory. No greater crisis could arise than that of the present time, when the world faces a pandemic the like of which has never before been contemplated, and defence against which is almost impossible using conventional pharmaceutical techniques. It is as inevitable as night follows day that the chemistry of life will be reformulated in terms of the electronic physics of life–a process long overdue.

We shall become less used to regarding the materials of which we are formed simply in material terms, but will learn at a popular level to recognize the energetic ingredients of matter, and the interchangeability of matter and energy, a concept initiated by Heisenberg and by Einstein in the first decades of this century, but never expanded in the direction of the human corpus, because, to use Einstein's own words 'the world is not yet ready' (2.62:49).

3. And so to Bed? Sleep and Geopathic Stress

I decided to give it one last try. I got up and slowly approached the place marked by my jacket, and again I felt the same apprehension. This time I made a strong effort to control myself. I sat down, and then knelt in order to lie face down, but I could not lie in spite of my will. I put my hands on the floor in front of me. My breathing accelerated; my stomach was upset. I had a clear sensation of panic, and fought not to run away. I thought Don Juan was perhaps watching me. Slowly I crawled back to the other spot and propped my back against the rock. I wanted to rest for a while to organize my thoughts, but I fell asleep.

I heard Don Juan talking and laughing above my head. I woke up. 'You have found the spot,' he said.

Carlos Casteneda, *The Teachings of Don Juan: A Yaqui Way of Knowledge*.

The previous chapter set out the mechanisms by which our brains control the repair of our bodies-'the healing brain' as Bob Ornstein rightly called it (3.1:195). I would now like to propose that this healing operation is carried out while we sleep–indeed that the very purpose of sleep is for cell repair–and once the mechanism of sleep is fully understood we can begin at last to see practical ways of protecting this important function against the hazards of electropollution.

The Study of Sleep

It occured to me only recently that, although sleep constitutes one third of a human lifespan, very little attention has been paid to it; we do not even know why we sleep at all. Even physiologists and psychologists have no certain idea why we sleep. In an epic book on the brain the famous scientist A.R. Luria devotes only two references to sleep, the first briefly mentioning that the brain's reticular formation, especially the posterior part of the hypothalamus, is a factor in determining the level of wakefulness, and the second reference suggesting that people fall asleep only because their constant flow of incoming information is curtailed. 'A normal person', says Luria, 'tolerates restricted contact with the outside world with great difficulty' (3.2:162). Yet for a third of every day we all shut down almost completely all our sensory and motor stimuli. Why? Luria, like everybody else, has no real answer.

Neil Carlson does rather better in his excellent textbook, *The Physiology of Behaviour*. He devotes an entire chapter to sleep, and in a brief review classifies its theories into two types; (a) that sleep is a period of restoration and repair, and (b) that it is an adaptive response to conserve energy. In support of the former view it seems that growth hormone production accelerates during the deepest states of sleep and that chemical neurotransmitters are also produced to cause the actual onset of sleep (3.3:39).

In 1976 Drucker-Colin's research team discovered an amazing thing. They perfused quantities of Ringer's solution into the brains of cats through a cannula tube fitted through their skulls. (Ringer's Solution is very similar to the fluids found in the brain's cavities.) They then reinjected the perfusate from the cats directly into the brains of other cats and watched what happened. If the perfused cats were sleeping at the time, then the injected cats also went to sleep. If they perfused fluid from cats which were awake into the brains of cats which were asleep, the injected cats immediately woke up (3.4:79). Of course, points out Carlson, these neurotransmitters may only be part of a process which is initiated by other (unknown) means.

This same cannula tube technique fortuitously supported the morphogenetic radiation hypothesis in another experiment monitoring protein synthesis. It was found that protein synthesis in the brain increases dramatically during sleep, with the highest increases occurring during what is called D-sleep (desynchronized–because the brain patterns become irregular). It was Hartmann who invented the terms D-sleep and S-sleep, when he found that the former type exhibited Desynchronized EEG records and latter Synchronized EEG records (3.5:117). It seems that in D-sleep all muscle tone is lost; the sleeper is 'out to the world'. Meanwhile, unencumbered by 'noise', the desynchronized EEG signals are propagating information to the body's cells.

During such times there is massive inhibition of alpha motor neurones, and yet at the same time there is enormous acceleration in brain activity. Cerebral blood flow and oxygen consumption are also remarkably increased. These effects were so surprising to early researchers into sleep, like Kleitman and Aserinsky, that they called this deep yet cerebrally-active sleep 'paradoxical', because it didn't make sense for the brain to be working overtime when there was nothing going on corporally (3.6:145).

Sleep Theory and Cerebral Morphogenetic Radiation (CMR)

Kleitman wrote in despair that: 'No one theory of sleep or wakefulness has so far been able to bring all the facts together in coherent form.' But the morphogenetic radiation hypothesis will neatly solve the problem. Periodically the brain must need to transmit repair instructions to those cells which through daily wear and tear have become damaged. It therefore shuts down the signals coming off all other kinds of activity and transmits and receives its unique signal to individual cells, telling them to initiate protein synthesis where necessary. It can do this more efficiently if the signal-noise ratio is favourable for transmissions, and an inert body helps considerably. It also helps to explain why all dividing cells seem to stop and wait at one point in their cell growth cycle (the G2 phase), as if waiting for a signal to proceed.

This phenomenon of massive inhibition of motor neurones enables the transmissions from the brain to be heard loud and clear, and the brain to receive more clearly the signals from individual cells. During D-sleep even tendon reflexes cannot be elicited–we become paralysed.

The D-sleep occurs in cycles of about 90 minutes, each containing a 20 to 30 minute period of D-sleep, so there will be five periods of D-sleep in an eight hour sleep, the most being accomplished early in the night (3.7:184). From this some sort of refractory period is indicated, and CMR hypothesis would presume that the protein synthesis instructions are being transmitted during D-sleep and executed in the refractory period. After a while, a new set of transmissions verifies that the original instructions have been executed; this will take less time than previously, so the period is shorter. In fact the periods will get progressively shorter. It should be easy to check quite simply by observing the protein synthesis in progress at a cellular level during the refractory periods, but not during the transmission periods of D-sleep. To date no such experiments have been carried out to my knowledge.

The length of sleep seems to diminish as any animal grows; babies do little more than sleep and eat, and their periods of sleep stay regular in time (for they are growing new cells constantly), whereas adults whose bodies have finished growing sleep much less, only enough to repair the cells they have lost during the day.

The orator Cicero (who probably didn't do a great deal of manual labour) could do with only four hours' sleep each night for many years. The curious finding of scientists is that sleeping pills actually cause insomnia according to William Dement since they suppress D-sleep (3.8:74). When medication is discontinued a large rebound in D-sleep is observed.

Early psychologists, like Donald Hebb writing in the late forties, were very hazy about the reasons for sleep. Hebb's review of previous theories suggests mainly that we go to sleep when we have nothing better to do (3.9:124). But some of the clues had already started to emerge even then; in 1941 Jasper was able to establish that the EEG in sleep showed a marked hypersynchronicity. Hebb himself noted that the waking centre in the caudal hypothalamus, when destroyed, led to continual sleep; and Nauta showed conclusively that the two parts of the hypothalamus (anterior and posterior) are the sleep centre (3.10:189). It had also been accepted that the slow EEG waves of the infant were similar to the same slow patterns of the sleep or coma of adults.

To the proponents of CMR the reasons for sleep are obvious; the brain is simply taking advantage of night and its lack of solar radiation—and at the same time shutting down as much extraneous internal electronic 'noise' as possible—in order to transmit and receive electromagnetic communications to the individual cells of its body, and instruct their repair. But no unified connecting principle has until now emerged. Let no one doubt that we lose our body's cells by the million during daily activity. Whether simply sitting in the sun, doing exercise, or working like Cicero at an office desk, some of our cells are dying, falling off or being altered or damaged in some way:

As Lyall Watson puts it:

'We shed sixty hairs a day. Scrape a thin sliver from a fingernail and you lose another ten thousand cells, stratified and compressed into a hard bony substance. On the outside of the body every touch, every breath of wind takes its toll, and on the inside conditions are just as rigorous. Every day the entire lining of the mouth is washed down the stomach and digested, and 70,000 million cells are scraped off the walls of the intestine by passing food' (3.11:259).

Repairing all these cells each day is a renovation task of enormous complexity. To do so the cells must each individually be given orders to initiate the necessary protein synthesis. We now know that one codon in the RNA macromolecule, and one only (Adenine-Uracil-Guanine), is at the start of a genetic message (3.12:108). But there is a growing amount of evidence to suggest that the brain is responsible for giving the necessary start instructions and that it does so most noticeably during paradoxical sleep. There is a definite link therefore between paradoxical sleep and organic protein synthesis.

Back in 1935 Adrian and Yamaguva conducted an early experiment (3.13:8), which showed that the newly-discovered alpha rhythms were not in phase over the whole scalp, but that there was a focus of activity on either side of the midline in an area about two inches laterally and two inches upwards from the inion. (This is no surprise in the context of CMR.) Alpha rhythms normally appear only when the eyes are closed, since during wakefulness they are replaced by the much faster beta rhythms. It seems that the alpha rhythms are the EEG's crude record of the signals to the cells, and that these are inhibited when the eyes are conducting their own electromagnetic signals to the occipital cortex.

That such a mechanism is at work is supported by the discovery by Peters and Jones (3.14:203) that there is a complete absence of callosal afferents to Brodmann Area 17, which (as far away as possible at the back of the brain so as not to interfere with the alpha signals), is in the occipital region of the cortex. In other words, these must be two distinct electromagnetic systems in the brain; the visual system and the CMR system. At last one begins to understand why the optic nerve is so insulated and separated from the rest of the brain, and why it crosses over, thus balancing the flow of ions.

Danguir and Nicolaidis provide further interesting evidence for the CMR hypothesis; they found that energy metabolism is directly connected to paradoxical sleep, during which the hormones involved in metabolism exhibit increased or nearly exclusive secretion. The same researchers also found a correlation between size of meal and the duration of paradoxical sleep, as if the brain was making use of the nutrients' availability in calling for protein synthesis (3.15:69). No wonder we often feel drowsy after a large meal; cell repair time is then at hand!

Conversely, the less the food intake, the less paradoxical sleep; in the extreme case of anorexia nervosa insomnia is a frequent symptom (3.16:65). One cannot escape the notion that this particularly difficult psychosomatic condition might be more successfully tackled from the position of encouraging paradoxical sleep in some way, which many modern drugs tend to inhibit. At the other extreme, hypersomnia as produced by the Kleine-Levin syndrome, induces hyperphagia (overeating). Mounier Jouvet, moreover, working in the sixties, confirmed that increased protein synthesis takes place during paradoxical sleep in post-natal ontogeny (3.16a:185). Finally, in unrelated studies also in the late sixties, Haider and Oswald found that amino-acid supply favoured specifically the time of paradoxical sleep (3.17:113).

All these experiments support the view of a direct connection between paradoxical sleep and protein synthesis. Strangely enough, the specific experiment which proved that motor and sensory muscular action is 'strikingly' suppressed during paradoxical sleep was only carried out in the 1970s (by Chase and Morales) (3.18:46). Without any general explanatory hypothesis of sleep however, their results only confirmed what we know anyway; that sometimes we are so asleep as to be 'dead to the world'!.

The CMR hypothesis allows us to construct a much more complete theory of sleep than any so far offered. It also gives us much more insight into the nature of the alpha rhythms themselves. Despite the fact that twenty two out of the seventy four Nobel Prize awards for biology so far this century were for neuroscientific discoveries, Olof Lippold in his book *The Origin of the Alpha Rhythm*, published in 1973, was forced to lament:

In no field of medical research has so much been written and so much work been carried out for such meagre advances in our understanding. ....The rhythmicity of the EEG as seen in alpha waves is just as great a puzzle as it ever was (3.19:28).

One bored but inventive American scientist called Rosenbloom even went so far as to connect up his own EEG records to convert his alpha rhythms via suitable electronic circuitry in order to record the world's longest musical composition; it lasted 72 hours, and he called it *'How much better if Plymouth rock had landed on the pilgrims'!*

CMR and Cot Death

If sleep is the way in which the brain gets its message across to the cells, it follows that infants, whose cells are growing at a prodigious rate, but whose brains are still not fully myelinated, are most vulnerable to damage from external electromagnetic fields (3.20:72). Babies sleep nearly all the time and their pattern of paradoxical sleep is maximal, with brief respites for feeding. I spent a few months investigating this prediction, and found that cot deaths in four London boroughs significantly correlated with proximity to important sources of electromagnetic energy, whereas controls showed no such correlation (3.21:51). Moreover, the nearer they were to such fields the earlier the hapless infants died. I also managed to measure the actual field intensities at the cotside in several dozen cases, and found that without exception the electromagnetic fields were many times normal levels. In these cases simply switching off the mains inevitably collapsed the fields, which had been due to unbalanced ground return currents, or such chronic sources as electrical hot water heating immersion systems.

My findings, predicted by CMR theory, caused quite a stir, and there was even a debate in Parliament on the matter (3.22:116).

Cot deaths in Britain kill nearly two thousand infants each year, and in the United States the figure is nearly ten thousand. In fact Sudden Infant Death Syndrome (SIDS) is the most important source of infant mortality today, and the most mysterious. The major facts about cot death all support the proposition that this tragic mortality results from chronic exposure to electromagnetic fields. There are many more deaths in the winter months, when electricity consumption is higher. A correlation with electric heating is often found (3.23:103) (3.24:210). It is also prevalent in poorer districts where electrical wiring systems are old and more liable to leakage. The deaths themselves occur just after a feed, when the cerebral morphogenetic radiation propagations are starting up. A German doctor Eckert had noticed many cot deaths near electric railways (3.25:85). When Gadsdon and Emery of Sheffield University carried out pathological inspection of hundreds of cot death infants' brains in 1976, they found that in more than half of them the poorly myelinated corpus callosic fibres had lost their fatty sheathing, which seemed to have melted, because the globules had re-coagulated around blood vessels in the lateral ventricles below (3.26:97;115). It takes a year to complete the process of myelination in human beings, the period in which many cot deaths occur. These infants' immature CMR systems in my opinion had been overburdened; in trying to increase propagation levels against the ambient 'noise' from surrounding EM fields, their systems had overheated and burned out. I have now developed a mesh cover, somewhat like a cat net, to guard against such hazards.

Geopathic Stress

Fortunately a new awareness of sleeping in the correct place is emerging. In the twenties George Lakhovsky, who had previously worked with Marconi, proposed his theory of natural cellular oscillation, and he did not take long to realize that chronic cellular dysfunction might be related to external radiative influences. In his case he was very concerned that cancers might be the result of sleeping in geographically dangerous spots. He thus devoted a whole chapter of his book *The Secret Life* to the influence of the underlying soil and water on human beings, and its contribution to the causation of cancer:

I propose showing how my researches in this direction have led me to establish that the nature of the soil modifies the field of cosmic waves on the earth's surface. This condition may be sufficient to cause in living creatures a cellular disequilibrium susceptible of giving rise to cancer (3.27:151).

Even before Lakhovsky, in 1869, Haviland, writing in the *Lancet*, suggested that 'The Thames and its tributaries are a vast cancer field' (3.28:120). Even before that a pioneer naturalist, Neree Boubée, informed the French Academy of Sciences that the cholera epidemic which was ravaging the country had a close relationship with the geological nature of the soil. Louis Pasteur, on his deathbed, is rumoured to have said: 'The microbe is nothing: it is the terrain after all,' and though he may not have been literally referring to the electric field, he finally recognized the importance of such environmental effects.

Lakhovsky emphasizes that the water running through a region carries the salts of that region. Where there is a deficiency of iodine, he points out, there is a higher prevalence of goitre, for example. Cholera breaks out on alluvial soil, whereas intermittent fevers show a preference for impermeable clays or marl. In April 1927 M. Stelys brought evidence to the Academy of Sciences in Paris that some soils were carcinogenic. Shortly after this, on 4 July, Lakhovsky unveiled his own explanation, based on Stelys' data, which had three components:

(1) demographic studies on the distribution of cancer mortality

(2) geographic studies of the soils on which cancers develop most freely

(3) physical consideration of the electronics of the soils in question.

Lakhovsky chose Paris as his area for research. He discovered that areas where cancer density is high, e.g. Auteuil (1.7 per thousand inhabitants), Javel (1.61), Grenelle (2.08), and St Lambert (1.57), rested on plastic clay. Areas where cancer density was low, such as Porte Dauphine (0.53), Champs Elysées (0.68), and La Muette (0.995), were on sand and sandy limestone. The figures were mixed where the soils were complex.

Lakhovsky believed that some soils absorbed natural cosmic radiation, thereby rendering it harmless, while others reflected and thereby concentrated its noxious effects on the people living on it. Millikan the physicist, he pointed out, had already proved that cosmic rays can penetrate 50 metres into the earth's surface.

In cities the influence of building materials such as stone, bricks, masonry, tar, asphalt, and paving stones need not be considered, for these eminently dielectric materials do not impede the propagation of waves (no wonder the blocks of granite above the Kings' Chamber of the great Pyramid were so massive!). With a wavelength of 16,000 metres penetration is effected to depth of 80 metres in an insulating soil (sand, limestone, etc.), whereas penetration only reaches a depth of two metres in sea water, which is a very good conductor.

It follows from Lakhovsky's view that the concentration of cosmic rays will be much higher above subterranean streams than where there are none, since the water streams will refract the radiation, though Lakhovsky himself did not draw this conclusion:

We have seen that a low incidence [of cancer] is found on the sands of Fontainebleau and Beauchamp, which consist of pure silicates (like the Pyramid of Gizeh), and as such are highly insulating. A low incidence is also observed on the sandstone of Beauchamp, the gravel of Geneva, and the friable sandstone of Bern, the slate, greiss, and granite of Nantes, and the gypsum of the north east of Paris. . . .The incidence is highest on the soils containing ores and collieries, such as St Etienne, Metz, and Nancy.

Lakhovsky goes on to explore the role of water, suggesting that since the conductivity of water is dependent on the salts within it (pure water being a very good insulator), the carcinogenic and re-radiative properties of water will be equally reflected by its conductivity:

Water does not play a part in the incidence of cancer except when its electric constants and the form of its volume (water beds, etc.), are of such a nature as to affect the field of cosmic radiation, which may break up the equilibrium of cellular oscillation. In the light of these facts we are in a position to realise why so many reputable writers have often drawn attention to the existence of 'cancer houses', and 'cancer districts'.

Among such reputable writers Lakhovsky would undoubtedly have included Dr H. Ernst Hartmann. Hartmann, a cancer research worker from the University of Heidelberg, drew attention to the fact that a medical observer, Chaton, had been impressed by the high incidence of cancer in the Ognon valley in France. In this valley the river flows in a bed of Jurassic formation abounding in plastic clay. Hartmann's work was recounted recently by Rolf Gordon, the tragic death of whose son at the age of 26 set him on a search to discover the causes of cancer. His research results, entitled *Are you sleeping in a safe place?*, proposed that cancer is caused by what he calls 'harmful earth rays'. Ernst Hartmann, he relates, claimed the existence of an electromagnetic grid ('the Hartmann net') running north and south, with N-S grid lines of 2 metres apart and E-W grid lines 3 metres apart, each line having a width of between 20 and 80 cm, depending on such variable factors as the full moon, environmental pollution, and sunspot activity (3.29:105).

Unfortunately, since it casts doubt on the whole concept, another kind of grid with the same harmful intersections was proposed by Drs Curry and Whitman, working at the West German Medical and Biochemical Institute (3.30:66). Curry's grid however has a pattern 3.5 metres square, and runs diagonally to the Hartmann net. Like the Hartmann net, the Curry grid lines are charged + or −, and where two similar charges intersect the disturbance is strong (3.31:6).

A third and different hypothesis comes from Dr Picard of Moulins in the French department of Allier, who, with the aid of the Institute of Geobiological Research of Lausanne, researched the homes of 42 persons whose medical history was known to him (but not to the Institute) (3.31a). The researcher mapped out the courses of all the subterranean water flows in the district where the patients lived. When the two sets of data were superimposed, a strong correlation could be observed between the 'cancer houses' and the underground streams, particularly where the two streams crossed at different subterranean levels.

It seems that animals, unlike man, have retained their ability to sense where such dangerous spots lie. As for the various grid hypotheses, one cannot be sure. The earth's magnetic field will also have lines of force which may be used for navigation. A most curious example of this occurred on 24 June 1988 in Europe. At 5.20 a.m. on that day, 3,000 pigeons were released from the town of Bourges in central France, and others totalling 40,000 from other cities in Europe, as part of one of the greatest annual international pigeon racing days. The pigeons were expected to arrive home the next day to many parts of the continent, Germany, Holland, Belgium, and Great Britain included. But by Saturday night only 283 out of a total of 5,000 British birds had returned safely. As *The Times* reported 'It was sheer bad luck that the event should have coincided with the largest solar flare for four years' (3.32:245).

The flare, which lasted 91 minutes, seriously disrupted the earth's magnetic field and caused the pigeons to lose their way. People I spoke to on that day all said they felt particularly tired and worn out, even though none of them at the time was aware of the solar flare which had caused the disturbance. How many times has your favourite house pet elected not to sleep in the place selected by you, but to eschew the basket for some other part of the room?

On a more academic level, an engineer called Leinert from Zurich placed white mice in a cage 3 metres long, part of which crossed one of these so-called 'earth rays'. He confirmed the mice would always make their nest in an area away from the rays. When the cages were reoriented by 180° the mice yet again re-formed their nests in a neutral area. When he re-located the mice in smaller cages whence they could not avoid the earth rays, they rapidly became ill and lost some 20 per cent of their body weight. They also developed tumour growths while their companions, which had remained in a neutral zone, stayed perfectly healthy (3.32a:94).

Rolf Gordon believes that anyone who persistently sleeps over earth rays will contract cancer. He is not alone; in 1929, the father of the theory, von Pohl, mapped out the small village of Vilsbiburg whose inhabitants numbered about 2,600. By comparing the addresses and sleeping places of 54 people who had died of cancer, he was able to establish that all these 'had slept in a strongly ray-infected place at the time the illness was diagnosed' (3.33:207). Von Pohl duplicated his findings at another 'cancer district', Dachau, subsequently chosen as a death camp. Dr Rambeau, president of the Chamber of Medicine at Marburg, could not fault his findings there, and numerous scientific experiments have since validated von Pohl's hypothesis (3.33a:235). Dr Hager of the Scientific Association of Medical Doctors, checked the houses of no less than 5,348 people who had died of cancer in the town of Stettin, and found in all cases that strong earth rays, as measured by a professional dowser, had crossed their homes. Three old people's homes were studied, in the first, where strong earth rays were found, 28 people had died of cancer; in the second, where only weak rays were found, only two had died and in the third, where no rays were found, none had died.

But the tests can be even more precise. Manfred Curry claimed to be able to forecast which part of the body would be most affected simply by knowing through which part of the bed the earth rays passed. The same results were recently reported by Ulrik, surveying 500 people with serious illness:

I am satisfied by the connection between illness and the location of earth rays. The patients I visited did not have to tell me where in their body they had pain, or where the illness was. I find out where the rays run in relation to their beds, and I calculate where the person has the cancer.

In West Germany on 16 February 1987 the Research Minister, Herr Reisenhuber, announced that the Government had allocated DM400,000 for research programme into the effects of 'harmful earth rays' on human beings and other organisms, to be headed by Professor Hildebert Wagner from the Institute of Pharmaceutical Biology at Munich University. Early in 1988 the British Central Electricity Generating Board also announced a £500,000 research programme into the possible biological effects of high tension low frequency power cables, electrical, and related appliances. Both the CEGB and the pharmaceutical industry have powerful reasons for not finding positive results and neither of these institutions are likely to pay much attention to the possibility of the existence of cerebral morphogenetic radiation. However, Rolf Gordon cites many individual cases where earth rays have caused not only cancer but other diseases, including rheumatic aches, headaches, arthritis, multiple sclerosis, epilepsy, hypothalamic tumours, asthma, depression, nose bleeding, tuberculosis, meningitis, heart attack, kidney stones, and even traffic accidents. All these except the last are diseases which seem particularly related to electromagnetic pollution in my view, and most concern the brain directly.

Though traffic accidents do not at first seem to fit the pattern, we shall see that once again it is the anomalies which prove to be most instructive. Robert Endros and Karl-Ernst Lotz were investigating why a high number of head-on collisions occurred on the same stretch of road for no apparent reason (3.34:88). The few survivors all reported a complete blackout prior to collision, and Rolf Gordon suggests that powerful earth rays may have upset the drivers' pH values, and thereby the endocrine system, sufficiently to cause the blackout. After further research he has invented a neutralizing device which uses microwave radiation to disperse harmful rays and can be used in any home. Other neutralizers are available for cars and factories, and it is claimed that the accident rate on one road was cut from 32 accidents in two years to one accident in one year after the installation of such a neutralizer.

All this seems a long way from the mechanisms of sleep with which we started this chapter. Before finally turning for home, so to speak, it is useful to relate Lakhovsky's insight in this area. He noticed that according to Dr Shannon in 1917, the city of Memphis uses artesian wells for its water supply and also has a low incidence of cancer. Shannon claimed that the comparative freedom from cancer was due to the absence of protozoan organisms in the water from these wells. But no one, pointed out Lakhovsky, has ever succeeded in linking cancer to the presence of protozoa in water, and he offers a different explanation. The Memphis water supply originates at the town, and is a mineral water possessing the same characteristics as the soil on which the inhabitants of Memphis sleep. Thus their body-cells have automatically the same electrical and chemical properties as the soil of their habitat, and consequently they are 'in resonance' with the local field of cosmic radiation.

Lakhovsky never came across the modern phenomenon of jet-lag, because rapid air travel at that time was meagre, and jet planes were not yet invented. But it cannot escape notice that his hypothesis would also explain jet-lag; air travellers quickly put themselves and their cells on to terrain with which, according to Lakhovsky, they would not be in natural resonance. In consequence, according to the CMR hypothesis, the signals from the brain are being corrupted temporarily.

Lakhovsky cites other towns, like Luxeuil in France, where there is an almost total absence of cancer, and where the inhabitants 'drink only the mineral water of the local spa establishment, obtained from the depths of the local soil'. At Chatel-Guyon there is a similar low incidence of cancer. 'Now it is known', states Lakhovsky, 'that the water supply of this town does not come from a distant source, but is derived from a local site, Mont Chaluset'. Moreoever, the water supply for Geneva (where cancer incidence was distinctly low at 0.5 per thousand in 1949) is drawn from the depths of the lake, presumably too low for cosmic radiation to have much effect.

If Lakhovsky's hypothesis is correct, hydrologists the world over will have to rethink the design of their civil engineering projects. And the manufacturers of mineral waters for which so much is claimed will have to rethink the potential harm to those drinking them far from the place of manufacture! Using a distant water supply at first glance seems actually to have a carcinogenic effect. In March 1927 Dr Simeray reported to the French Academy of Sciences that the population of an entire village had been free of cancer so long as they used the local wells. When the local authorities changed to a water supply from outside the locality, a series of cancer cases occurred in the village. Whether the same applies to other ingested liquids, like milk, orange juice and so on has not to my knowledge been researched–but these all have a high water content. City-dwellers, take note! However, I prefer to think that the flow of water along the pipes, which is known to create an electromagnetic field, is the more likely cause of cancers, as a later chapter will show.

Lakhovsky's contribution to the theory of sleep is rather bizarre, but if his concepts prove correct, it is a lifesaver:

The disadvantages of modern water supplies might be overcome in cities by sinking artesian wells such as those existing in Paris in the Place Lamaritine, the Avenue de Breteuil, and the Bois de Boulogne. As for the new well of the Rue Blomet, it were infinitely better if this were used for household purposes rather than for a swimming pool. When local living conditions are exceptionally bad or variable, it is possible to re-establish or rather 'tune up' the electrical constants of the cell by means of appropriate substances in harmony with the physical and chemical nature of the soil of the habitat.

(He means, to eat vegetables grown locally to where we live; worshippers of organically grown fresh foods unfortunately buy and eat them far from the place they were grown, which rather defeats the object!)

These substances could be administered by hyperdermic injection, or preferably by the oral route. At night time the sleeper might be connected with the soil by means of an appropriate earth connection, and in the daytime footwear might be used to hold a metallic plate in the sole or heel, thus establishing contact between the foot and the soil.

In the majority of cases, it seems more rational and efficacious to resort to electrical methods such as filtration of the field of cosmic waves in the immediate vicinity (3.35:151).

Four decades or so after Lakhovsky's death in 1943, new French proponents of geopathic stress, as it is now called, have come forward: Jacques la Maya in 1983 produced his voluminous *La Medicine de l'Habitat* (3.36:170), Jean de la Foye, who died in 1982, wrote *Ondes de vie, Ondes de mort* (3.37:90), and Lucien Roujon, a bioelectrician working as a researcher at the department of molecular biology of L'Université de Paris-Sud, Orsay, wrote *L'Energie Microvibratoire et la Vie* (3.38:218).

Roujon proposes that three influences determine organic health: the pH factor (this is the hydrogen potential, commonly used as a measure of acidity-alkalinity); rH (the electric factor, measuring the level of oxidation-reduction), and r (resistance to an electric current, the insulation factor). These summed together indicate how resistant any biological substance is to disease, says Roujon. By locating various foods, vegetables, fruits, drinks, and diseases and their vaccines on a graph with pH and rH as parameters, one researcher from Toulon, L.C. Vincent, shows how the vaccines tend to balance the diseases towards perfect health, which lies at the centre of the quadrants. He calls his graph a bioelectronigram.

Those suffering in ignorance from harmful earth rays tend to rationalize the reasons for their dis-ease; if we are persistently tired, lacking appetite, moody, or depressed we tend to explain this by excuses connected with events in our lives. Von Pohl asserts that certain plants and animals actually thrive on harmful earth rays, while other creatures avoid them; this is one way of detecting them. I happened recently to live in a fifteenth-century chateau in the Loire Valley where some of the plants thrived while others did not. A nectarine in the moated court-yard was always sickly and eventually we dug it up. According to Rolf Gordon nectarines and peaches thrive on harmful earth rays. On the other hand von Pohl says: 'Peaches are the most sensitive towards earth currents, and will not grow at all on a cross', so, like astrology, there seems always to be an escape route for the interpreter.

At the chateau when they came down to breakfast, guests would always exclaim how well they had slept the night before. Since the walls were a metre thick it is likely that some of the artificial EM fields at RF frequencies were at least deflected. Curiously too, the sheep of the place would never stray far, even though it was easy for them to wander on to the nearby road. I often wondered whether the original site engineers had chosen the spot deliberately, since although it guarded a tributary of the Loire, it was sited over half a kilometre from the nearest river bank. Set in open countryside with plenty of better locations further from human habitation available for nest building, the spot was nevertheless a favourite for nesting birds such as hawks and waterfowl, and on one occasion we had to wait a fortnight while a young brood of mallards hatched out. Only when I saw them setting off in a fluffy line behind their proud mum for the nearby river was I at last able to cut down the undergrowth in preparation for a new lawn.

4. How to Handle Power Lines

All power corrupts. *Lord Acton*

Having briefly proposed some of the biological mechanisms at work when electropollution affects us, I now turn to the practical aspects of what we can do to protect ourselves. This chapter looks at power lines, by which I mean not only the wires strung out across the country which transmit electricity from power generating plants to the end user, but also the distribution lines, down to 11,000 volts, before they are transformed down again to the voltages (415 and 240) which we use in our homes. All these are known as power frequency lines, because they alternate at a frequency of fifty cycles per second (50 Hertz). In America and some other countries they alternate at 60 Hertz, which is slightly further away from the frequency at which our brains themselves oscillate all the time we are alive.

In 1985 Rosalie Bertell chronicled the dangers of low-level radiation–and how they have been ignored for so long–in her book *No Immediate Danger* (4.1:26).

If scientists had realized early enough how harmful non-ionizing electromagnetic energy fields can be to people who live near them, even at very weak strengths, the electricity generating utilities might have rethought the design of the National Grid, and, like some Scandinavian countries, elected to place their generating stations close to the cities and industries which consume their power so as to reduce the number of high voltage lines necessary to deliver the electricity. However, fear of the effects of nuclear emission, cost considerations (urban and suburban land is expensive), as well as the sources of fuels like coal, or coolants like water, have dictated that the power stations, at least the nuclear ones, are built on remote coasts like Dounreay in Caithness and Sellafield in Cumbria.

Let us hope that new power stations at any rate will be established in proximity to the industries and the populations they serve, and adequately protected by Faradic shielding. Too late now! We must make the best of a bad job.

Research into the Effects of Power Lines

It is estimated that over 80,000 citizens in the United Kingdom, and more than a quarter of a million Americans live close to high voltage power lines. Valerie Beral, one of Britain's top epidemiologists working at the London School of Hygiene and Tropical Medicine, has said that it might be as much as one per cent of the population in the United Kingdom, which is about half a million people (4.2:55). My advice to them all is, move house! This is not so drastic a remedy as all that, because people on average move home every seven years anyway these days, as a result of changing job, getting married, divorce, retirement and so on–a measure of the mobility of our modern society.

In 1989 in Vancouver the British Columbia BC Hydro utility announced that it was prepared, upon request, to pay a fair market price to landowners concerned about electromagnetic emissions from its new 230 kV line. The 90-mile Dunsmuir-Gold River power line was built on an existing right of way which already had two 138 kV power lines, but the new line, which will serve a pulp and paper mill, according to Louis Slesin's *Microwave News*, came on stream in July 1989.

When the company sent letters to those whose property was within 50 metres of the edge of the existing right of way they received acceptances from 153 owners, 90 per cent of those approached. Only six people decided to stay. Those contemplating an investment in PowerGen should bear that in mind! Compensation of say £100,000 to each of the 80,000 members of the powerline residents' club of Great Britain would cost an unthinkable £8,000 million, and probably bring down the Government!

Nancy Wertheimer and the Denver Study

It may become illegal eventually for developers to build new houses near existing power lines anyway, in which case your house or flat might suffer a dramatic fall in value, so get out while the going is good! Ideally I'd like to see a Certificate of Electromagnetic Field Safety issued for every house put up for sale. There is a good deal of evidence to support your move. It really began in 1974 when Nancy Wertheimer, from Denver, Colorado began trying to see if she could unearth any environmental factor behind childhood leukaemia, a tragic condition which unaccountably causes the death of many hundreds of children in the United Kingdom each year. Paul Brodeur works as a columnist for the *New Yorker* magazine (which pioneered Rachel Carson's classic work on chemical pollution *Silent Spring* in the sixties (4.3:40)). His brilliant book *The Zapping of America* is one of the earliest and best accounts of electropollution. In a series of *New Yorker* articles, he described Nancy Wertheimer's moment of insight:

'It was on my third or fourth trip', she told him. 'I had stopped at one of the birth addresses — an old frame house on the edge of the warehouse district in downtown Denver — and I walked into an alley behind it, which ran between the back-yards of two rows of similar houses.

It was a mixed neighbourhood — black, white, and Mexican — and many of the houses were run down. Like most of the back yard alleys in the Denver area, this one was paved, so that trucks could get through it to pick up garbage and trash, and it contained a number of poles strung with electrical wires and telephone lines. When I looked up I noticed an electrical transformer on a power pole behind the house I had stopped to examine. The transformer was black and cylindrical, and it was attached to the pole above a crossbar that supported several electrical wires, and when I saw it I said to myself, Hey, haven't I been seeing a lot of these lately?' (4.4:35).

As she went round the houses where the victims had lived it gradually dawned on her that many were in close proximity to the primary and secondary wiring configurations which transformed voltages of 7,600 volts down to 240 volts for domestic consumption. As in the United Kingdom, such transformers are often attached to poles which can also carry telephone lines and other cables. Only when she read a news article suggesting that power lines might be a danger to health (possibly based on Dr Louise Young's 1973 book *Power over People* (4.5:271), or *Pollution by Electrical Transmission* (4.6:270), which came out the following year), did she begin in earnest to investigate their connection with childhood leukaemia.

She asked for technical assistance from a physicist, Ed Leeper, who explained the intricacies of electromagnetism and designed a field measuring meter for her. Together, at their own cost, the team documented a total of 344 cases of children suffering from cancer, twice as many as might have been expected to live near the wiring configurations. The results were not finally published until 1979, when they appeared, after scrutiny by other scientists, in one of the country's most authorative journals, *The American Journal of Epidemiology* (4.7:263). Wertheimer and Leeper said:

At these points the voltage has been stepped down and 'transformed' into current...it was particularly close to those transforming points that were over-represented among our cancer cases.

She went on to say (after conceding that magnetic fields are usually cancelled in normal wiring, where the return current tends to balance the supply current) that:

Such cancellation is imperfect in the vicinity of many dwellings because the wires are often separated in space, and more importantly because some of the return current does not flow through the wire at all, but returns instead through the plumbing system to which most electrical systems are grounded at each house.

Whereas the fields from a domestic appliance are intermittent (we don't use the vacuum cleaner all day, or continually boil a kettle) and fall off rapidly with distance, the fields from a high voltage power line are continual and pervasive.

British Research

A British example of this can be found in the density of transmission lines emerging from power generating plants. Both at Dounreay and at Sellafield (the latter embracing the original Calder Hall, the first Magnox nuclear generating plant, opened by the Queen in 1956), unusually large clusters of childhood leukaemias have been noted. One group of families has gone so far as to sue British Nuclear Fuels plc on the basis that their radioactive emissions from the re-processing plant are responsible for some, at least, of the 32 cancer cases recorded in an independent Advisory Group Report chaired by Sir Douglas Black in 1984. But neither his enquiry, nor three separate enquiries conducted by COMARE (a committee appointed by the Government to monitor environmental radiation), found sufficient ionizing radiation in the region to justify that possibility (4.8:30) (4.9:58) (4.10:57) (4.10a:62). Even so, no one is denying that the clusters exist, and that they are unusual.

In 1987, Dr Leslie Hawkins of Surrey University's department of Occupational Health, and a colleague, had suggested that the cancers might be the result not of radionuclides but from Calder Hall's emerging transmission lines. (4.11:122;242). Following up this notion I visited the Sellafield area in September 1989 and correlated such leukaemia cases as I could find against the course of the power lines. My task was not made easy when, on arrival, I found that Whitehaven's library file on the subject had gone missing, and the Environmental Health Officer at Sellafield, though providing much publicly available data, would not or could not release to me the individual addresses of the cases, nor even the annual amount of electricity delivered to the Grid from Calder Hall (which is not administered by the CEGB, so does not appear in the Electricity Council's annual handbook of statistics).

I found that the lymphocytic leukaemia cases were never far from the lines. Another group of cancers, the myeloids, found more often among older people, seemed to cluster not round the lines, but round radio and television masts and on high ground near the tops of hills. Wertheimer had already pointed the finger at the effect of low-level electromagnetic fields on organic life, drawing attention to one early experiment on slime moulds which found that cell division (mitosis) was inhibited by such fields (4.12:104), and that navy and other research had found similar effects could be brought about in seedlings, chicks, rats, and embryonic tissue cells. Later Dan Lyle and his team at Loma Linda, California were to prove her right even so far as human T-cells were concerned (4.13:163). She had also detected that there was an increased risk of cancer among people working in electrical occupations (4.13a:213).

Not unnaturally the United States power utilities were quick to respond to this potential threat to their operations. They set in motion research aiming to replicate, and hopefully thereby to disprove, the possibility that their wires were causing cancer in children. By 1980 another survey had appeared, from J.P.

Fulton and S. Cobb, analysing childhood leukaemia in Rhode Island and its relation to wiring configurations. They found no statistical relationship (4.14:96). As Wertheimer and Leeper pointed out later, this was not surprising: in studying their 110 cases of childhood leukaemia Fulton and his team had restricted the reference distance to 46 metres from lines. When Wertheimer and Leeper corrected this data for a bias in the controls towards urban residence (and thus greater exposure to high current configurations), the elevated risk was confirmed even in the Fulton study. Fulton had also made an error in assuming that the fields attenuate with the inverse square of the distance instead of its simple reciprocal. Accordingly the cases and controls were redistributed towards the lower dose categories, and thus appeared non-significant (4.15:264).

The Wertheimer-Leeper criteria for high wiring configurations were:

(a) less than 130 feet from one three-phase-primary (large gauge) or more than six primaries (small gauge)

(b) less than 65 feet from three-to-five-secondaries (before any service drop)

(c) less than thirty feet from a first-span-secondary (before any service drop)

A service drop occurs after the current had been transformed downwards from 7,600 volts (primaries) with no loss of voltage from the 240 volts (secondaries). In the United Kingdom the nearest equivalent is where an 11,000 volt or 6,600 volt line is transformed down to 240 volts. These transformers too are often hung on poles near houses.

I have been called to a farm in Somerset where a woman who had lived there for 13 years had been experiencing health problems for 11 of them. (The farmhouse was right out in the country, with no near neighbours, and the transformer was only about twenty metres from the property.) She suffered chronic asthenia, severe loss of weight, colonic complaints, candidiasis, and sleep apnoea; all-too familiar symptoms of an immune deficit. The electric fields inside the house were fairly normal on the ground floor, but upstairs in the bedrooms they were abnormally high, grading down from 70 volts per metre (V/m) at the end nearest the transformer to 30 V/m furthest away. As if from instinct, the woman was now sleeping in the furthest bedroom, but electric fields here were still three times the residential norm.

This is a typical situation; doctors were baffled. There was no response to treatment by their drugs, such as Caprosin, and Prosymbioflor, and an expensive oxygen machine did not have any real impact on the chronic ear nose and throat problems which the lady had experienced for years. The illness was undoubtedly exacerbated by a road accident a few years back. My advice was to ask the Electricity Board to move the transformer to a more distant pole, which would have inconvenienced no one. To date this has not been done, and the Board claims that the fields are less than those present in the house.

One of Wertheimer's detractors was Dr Morton Wheeler (funded by the United States Department of Energy at Rochester University), who claimed that the magnetic fields from the lines were lower than those propagated by a 150 Watt lightbulb. She pointed out in reply that this would only be the case if the lines are perfectly balanced. Indeed, I have measured fields inside bedrooms well over 70 V/m (the normal strength in any home is about 10 V/m), which disappear the moment that the mains electricity is turned off, proving that it is the domestic internal wiring circuit which causes the high field.

But the lines which Wertheimer and Dr Leeper researched are not really the high-voltage lines we associate with pylons; they carry only a fraction of the current transmitted by the huge 132 kV, 275 kV, and 400 kV lines criss-crossing Britain (4.16:86), of which there are 19,000, 4,000, and 9,800 circuit kilometres respectively. In America these lines can be as high as 765,000 volts. The fields induced by the latter types can be measured well over 100 metres away from their midspans. For example, the electric fields from the twin 132kV lines emerging from the north of the Calder Hall power station exceeded 50 V/m at a distance of 110 metres and at a height of 1.5 metres from the ground (the height of a standing person's head). These lines therefore irradiate the edges of the villages of Beckermet, Moorhouses, and Cleator Moor to the north of Sellafield, and the southern 132kV line irradiates part of Drigg to the south. All these villages have witnessed cases of childhood leukaemia in recent years (4.17:30).

Lennart Tomenius, a medical officer in Stockholm, Sweden, also researched in this area. Since in Stockholm the residents are fortunate in having most of their electricity delivered via buried cables (which may explain why cot deaths are so low), Tomenius chose to measure the actual field strengths at the door of the residences, and to note any kind of electrical transmission structure within 150 metres of them. After analyzing his data, which embraced some 2,098 homes of under eighteen-year-olds, he too found that twice as many children who lived near 200 kV lines developed cancer compared to the control children (4.18:247).

To chronicle in detail the subsequent arguments between the power authorities and these epidemiologists might begin to bore the reader (4.18a:219) (4.18b:220). Suffice it to say that the design of many of the utility-funded studies somehow seemed to avoid the issue; Myers, for example, in 1985, reported a negative result, yet conceded that up to 15 per cent of the cases had been excluded. He also restricted his data to cases within 100 metres of the lines, when it is clear that abnormal fields are found further away than this (4.19:186).

Another study by McDowall looked at 7,920 cases of death of persons living within 50 metres of electric transmission facilities in East Anglia at the time of the 1971 census. The study excluded underground cables, which is a pity, because the Eastern Area Board has more underground cables than any other by almost a factor of two. It also took the residence date at the time of the census rather than the time of birth, and then tried to see whether more than a proportionate number of people had subsequently died. Only 19 of the many thousands in the sample were living within 30 metres of an overhead powerline, and the majority were living near to the metal-encased substations transforming current downwards, which thus had their own built-in faradic protection. Not surprisingly McDowall's study found no positive results, and it was only later when he examined the fate of people working in electrical occupations that the correlation between ill health and electricity re-emerged (4.20:171) (4.21:172). Commenting on this study, Valerie Beral rightly points out:

The extent of migration of study subjects away from their 1971 address during the 12-year follow up is unknown; this lessens the certainty of their exposure classification, but it should be noted that the effect of this would be to bias the results towards the null hypothesis (4.22.55).

Beral mentions another survey of leukaemia incidence close to power lines carried out in four London boroughs and relating to cases between 1965 and 1980, which again pointed to an increased incidence of lymphatic leukaemias (though not of other types) among people who lived less than 25 metres from a power line, the relative risk being 1.76, falling to 1.45 at 100 metres. The risk from living near a substation was less, at only 1.3, but all these figures really mean that there is a hazard to health, since the normal relative risk would only be 1.0. (4.23:54).

No one seems to have taken much notice of these findings, perhaps because they were published in a rather restricted set of conference papers. What I draw the reader's attention to, though, is that it is always the lymphatic type of leukaemia which shows raised incidence. In other words, something has happened to the lymphocyte balance in the human bloodstream as a result of exposure to electromagnetic fields. The significance of this will become apparent when I talk about AIDS later.

The sad fact is that the CEGB's approach appears to be one of minimal research into health effects. Despite a budget of £100 million to promote its good image to the public in the run up to privatization, only some £500,000 has been put up by them for external research into health hazards, and even this was a grudging and belated response to a court directive following the Innsworth enquiry, a decade before. This compares with US $5 million spent on the New York State Power Lines Project, researching the biohazards of electricity transmission in the United States, some US $6.1 million by the American EPRI between 1985 and 1987, and a further US $16.7 million budgeted by them between 1988 and 1990.

What Can You Do?

Given this nonchalance by the authorities, what can ordinary people, do about it on a practical scale? To start with, here are my own guidelines for how near to a power line one might safely live:

Type of Line	Minimum Safe Distance (metres)	
	Sleeping	Long-term working (more than 4 hours)
400 k/V	250	100
275 k/V	200	150

Type of Line	*Minimum Safe Distance (metres)*	
	Sleeping	*Long-term working (more than 4 hours)*
132 k/V	150	75
11 k/V	50	25
Transformers (11 k/V 240 V)	50	25
Substations (metal clad)	100	50

As a rough guide, never allow yourself to sleep in an electric field exceeding 10 volts per metre. Young children should ideally not be allowed to sleep in any electric field at all, because of its effect on their cell division and long-term genetic structure. Since electricity has only been used this century the long-term hazards may not yet have become apparent, and there may simply be no safe level.

If you can't comply with these, admittedly somewhat arbitrary, criteria (based on cases I have personally encountered plus a careful reading of published scientific results), then there are still protective measures you can adopt. First, you can screen the worst effects of electric fields by growing trees between your house and the lines. Remember that the worst fields are found at the mid-span rather than nearest the pylon itself. Fast growing conifers like *Cypressus Leylandii* are good for this purpose, though shrubs can also be used to stop the effects at low level.

Most people sleep in the upper rooms of their house, where fields are generally higher, in such cases I suggest that you use rooms on the side of the house furthest from the line as bedrooms, or if possible sleep downstairs. The most effective measure you could take would be to construct a Faraday Cage in the bedroom! This would consist of a copper sheeting all round the walls and ceiling, and is clearly not a viable option for most people. You can use a metallic-coated material it as a cover for the bed, however. I have developed such a material, and in the course of doing so discovered that not many conventional metal threaded materials were any good. The windows can be protected by metallic mesh, which thereby still admits light and air. As a makeshift, aluminium kitchen foil under the mattress will protect you from fields beneath the bed, but gets exhausted after about a week, and will need replacing; unless you have earthed it well to the ground outside by means of a copper wire and piece of metal tubing hammered two feet into the soil. Many older models of mains give off quite high fields, and should be replaced by new metal-clad (*not* plastic clad) mains and fuse boxes.

Unfortunately, the Electricity Board now use radio telemetry, and thus prefer plastic to metal clad metering boxes. Before resorting to such desperate measures, it is a good idea to test the concept, by switching off the mains electricity at night for a few days to see if you sleep and feel better for it.

The human brain will adjust as best it can to any adverse fields, and if you go away, on holiday for example, your brain will be unprepared for the adverse conditions on your return. Perhaps you have noticed that you often catch a cold just after coming back from holiday, or that the first night away from home you simply cannot seem to get to sleep. Psychologists conducting sleep experiments are well aware of this 'first night' phenomenon. ME sufferers often report the first onset of the condition just after returning from a foreign holiday.

Stopping the effects of magnetic fields is much more difficult than stopping electric fields. They attenuate much more slowly and can get through most substances, especially water, of which human beings are mainly composed. The magnetic field from a power line comes out perpendicular to the line. By sleeping at right angles to the line therefore you are minimizing your exposure to its lines of force. Metals concentrate these force lines, so avoid sleeping with any metal objects near you, e.g. the radiators along the wall, or metal bedsprings. It is not so much the steady field which causes the damage as the changes which result from varying calls on the electricity supply down the nearby power line. The brain seems able to handle a steady field much more easily. That is why a single measurement of a magnetic field tells one little about whether the location is dangerous; it is the extent to which the field varies during a period of time that matters. Similarly if you are continually crossing lines of magnetic force in a geopathic stress zone, the same effect occurs.

The effect of an ordinary magnet on your colour television picture is likely to be pretty drastic if you bring it close to the screen. Not only will it distort the image but the people on it will end up with green faces for a few hours! It is possible that the brain too is affected, even by static magnets such as are often incorporated in the electric motors. If someone turns on a single electric light switch somewhere in the house, your brain will register the change and alter its EEG rhythms accordingly (4.24:19).

In one experiment, with identical twins, it was found that the brain of one twin thus affected caused a change in the EEG record of the other, even when they were in different rooms (4.25:80). Hans Berger would have been interested to know that! It is my personal opinion that this interpersonal communication might form the basis, not only of telepathy, but also of a mechanism for the transmission of progressive genetic ideas. If so, the phenomenon would explain how evolution proceeds much faster than traditional Darwinian natural selection predicts. The notion that a species, having learnt something, can then transmit their learning to others of their kind, is being developed by Rupert Sheldrake, a biologist who caused a rumpus in scientific circles when he first published his idea, called Morphic Resonance (4.26:224).

Given such amazing biological sensitivity to electromagnetic radiation, it is not unlikely that even the safe distances I have suggested may be too near. Only more research programmes and time will tell. Meanwhile, it's perhaps better to be safe than sorry.

We are beginning to know how to recognize the initial signs of electrostress, thanks to a pioneering set of studies by a Birmingham doctor, Stephen Perry. Once again Perry had to pay for the cost of his research out of his own pocket. He examined the addresses of some 600 suicides reported in the Birmingham area, and found that in homes where the magnetic field as measured at the front door was relatively high (above 1mG), the relative risk of depressive illness was elevated (at 1.5) (4.27:200). I found a similar effect in the housing estate at Fernhill, Mid Glamorgan, and in discussing this with several residents who had not been there long, was told that since arriving on the estate they had become depressed, and would be glad to leave. There had also been two suicides, not to mention many acts of violence and at least one murder there. A Yorkshire television programme on stress found similar symptoms among VDU-operating credit managers at a large finance company, Lombard North Central (4.28:136).

The lesson is clear; if you find yourself inexplicably depressed or suicidal and live near a power line, it's time to get the hell out, before some tragedy occurs. Perry's results, which were found to occur even when the lines were buried and the subjects were unaware of their existence, were confirmed by Italian and Russian studies, which also identified cardiovascular effects (4.29:276). I was able to demonstrate this to a resident of Fernhill who lived very near the 132 kV line by suggesting that he turned off the mains electricity. He immediately reported, literally within seconds, that the pressure in his neck had subsided, though at that time he had no idea of the research which had already found that blood pressure is affected by electromagnetic fields. So high blood pressure (which can also lead to aggression), is another pointer. If left, you may, like the inhabitants of Fishpond in Dorset where a power line stretched through their village, also suffer from hypertension, epilepsy, dizziness, and other heart conditions (4.30:231).

Make no mistake, despite the official denials, the scientific evidence, let alone my own personal investigations, have convinced me that these risks are real and seriously under-reported. It is hardly surprising that the electricity utilities deny it. After all, as Mandy Rice-Davies once said, 'They would say that, wouldn't they?'

5. Radio, Television, and Microwaves: Something in the Air

You shall hear their lightest tone
Stealing through your walls of stone.

Alfred Noyes, quoted in *The BBC Handbook*, 1928

Radio Waves

In November 1920, only 25 years after Heinrich Hertz first discovered radio waves, and four years after Albert Abrams had published his *New Concepts in Diagnosis and Treatment*, based on the same principles, the world's first commercial broadcasting station came on air in Pittsburgh, in the United States. Almost exactly two years later, on 22 November 1922, Sir John Charles Walsham Reith set in motion simultaneously from London, Birmingham, and Manchester, the first 'wireless' transmissions of the British Broadcasting Company Ltd.

The new medium caught on like wildfire. By the end of 1927 there were over 2.3 million radio licences issued in the United Kingdom (fee ten shillings), and thanks to the new Daventry transmitter, broadcasting at 610 kilocycles and 491.18 metres with a power of 30 kW, about 80 per cent of the British public, some 39 million people, could hear its programmes (5.1:215).

It immediately became apparent that an international code of regulation would be necessary to avoid stations jamming each other.

The *BBC Handbook* of 1928 stated:

The development of broadcasting in foreign countries as well as in Britain has shown that there is a limit to the number of wavelengths or channels that may be employed. This is due to the fact that even a moderately low powered station in Europe may spoil the performance of another similar station in Europe, however far away, unless the frequencies are separated by more than a definite amount.

That such European radio transmissions may also impact on biological systems was realized by very few; only by Dr W.E. Boyd, whose name is long forgotten; by Georges Lakhovsky, a Belgian scientist researching into the effects of electrical radiations on insects and human health, scarcely any better remembered; and possibly by Tesla, who had already started dreaming about a death-ray, and the concept of transmitting electrical power wirelessly. Tesla is mainly remembered today by the use of his name as a measure of magnetic field strength.

To the ordinary listener, radio was a miraculous phenomenon—even to sober scientists like Alexander Fleming (later to stumble upon the antibiotic properties of penicillin). In a letter home from his Arctic survey ship, written in a peaceful Norwegian fjord in 1924, he marvelled that he could listen to the music of a dance band as it played in London's Savoy Hill studio.

For Dr Boyd, on the other hand, the new 'wireless' waves were a nuisance. He found they were interfering with his delicate experiments as he tried to develop the diagnostic techniques based on 'electronic reactions' pioneered by Dr Albert Abrams of San Francisco (5.2:3). In desperation he built an enormous Faraday Cage in his testing room, whose walls, ceiling and floor were entirely clad in 30 gauge copper sheeting, and only then was he able to diagnose accurately the different substances in a phial at a distance, simply by means of changes in a human subject's muscular tone.

Abrams had already shown, before radio began to fog the possibility, that if a human being was brought into near contact with a specimen of diseased organic tissue, or even near to x-rays, the stomach muscles registered the change. With practice one could even distinguish between different morbific substances. Here, therefore, was a completely new method of diagnosing disease—unheard of in medicine—based on radio waves.

More surprisingly, Abrams was able, by generating radio waves of a specific frequency with an instrument which he called an oscilloclast, to correct the disease, inverting the signal to cause its cancellation. He had begun to teach these diagnostic and curative techniques at his laboratory, with such success that thousands of people, some of them by no means medically-qualified practitioners, were beginning to use his methods. Viewing the possibility of this upstart competition in their traditional business with no little concern, US doctors began a sustained attack on Abrams, which may well have contributed to his early death.

Dowsers undoubtedly sense the change in their own muscle tone, as evidenced by the twitch of their willow divining rod, to detect the electromagnetic fields given off by the passage of subterranean water as they pass over it. Such is the sensitivity of the human brain to the propagated waves of electromagnetic energy which we loosely call radio waves.

The public, nevertheless, was oblivious to these medical experiments with their sombre implications, and, along with their doctors, laughed Abrams and his 'Black Box' into the grave. By 1929 there was a total of 200 broadcasting stations in Europe, including 11 in Russia, having a combined power of 600 kilowatts. Two years later, according to R.N. Vyvyan's history of Marconi and the early days of wireless (5.3:253), the number of stations had increased to 261, including 46 in Russia, and their combined power had swollen to 2,860 kilowatts. One reason for this increase in power was to increase the range of government propaganda:

Every increase in power of broadcasting transmitters in one country leads to a similar or larger increase in the power of the stations of its neighbours, and so the evil goes on...until by the severe pressure of the body of its listeners the various governments will be forced to take action to reduce the interference.

Of course Vyvyan was not concerned with the biological effects, simply the interference with listening pleasure. At that time short wave transmissions were still experimental, but it was already recognized that they could travel great distances by bouncing off the 'Heaviside layer'.

The Electromagnetic Spectrum

Radio frequencies occupy quite a wide section of the electromagnetic spectrum, lying above the power frequencies and below those of ultraviolet and visible light. In that region we classify the frequencies as follows:

Type of Wave	Frequency Region	Wavelength Band
a) Radio		
long	50 KHz	10^3-10^4 m
medium	1 MHz	10^2-10^3 m
short	10 MHz	10^1-10^2 m
b) Television		
VHF	100 MHz	1-10m
UHF	1 GHz	10cm-1m

c) Microwave (300 to 300,000 MHz; equivalent to one metre down to one millimetre)

SHF	10 GHz	1-10cm
EHF	100 GHz	up to 1 cm

Microwave ovens radiate at 2,450 MHz (2.45 GHz), which is why so many scientific studies use that frequency; they can get the instrumentation inexpensively!

Just as there is little doubt that power frequency electromagnetic fields (*below* radio frequencies on the EM spectrum) induce biological effects, so there is ample evidence that the visible and ultraviolet frequencies (*above* radio frequencies on the EM spectrum) can also cause pernicious damage to cells. Professor Ronald Marks of the University of Wales medical college, a specialist in dermatology, notes the effects of visible light, sunlight, on the human body:

The evidence that persistent exposure to the sun is a potent cause of skin cancer is overwhelming...there have been many experiments which tend to confirm that the UVR in the sun's spectrum has a cancer producing effect. Most of these experiments have involved shining artificially-produced UVR on laboratory mice over periods of several weeks or months. Although their skin only has broad similarities to human skin, there is no doubt that the effects of this type of irradiation bear an uncanny resemblance to the effects of sun on human skin. Using this approach, researchers have been able to incriminate

UVB as the major waveband responsible for the carcinogenic effects of solar exposure...UVB seems to damage the cells' genetic material — its deoxyribonucleic acid, or DNA.

Another important effect of UVB may be on the skin's capacity to protect itself immunologically via the Langerhans cells...if the immune defences are compromised by the action of UVB in depleting the skin of Langerhans cells, any cancer cells that emerge in the skin may be allowed to flourish (5.4:168).

He goes on to say that up to 20 per cent of patients who have kidney transplants (an operation involving deliberate immuno-suppression) are affected by skin cancers within a decade of the operation. In a concluding statement, on which I shall enlarge later in this chapter, Marks simply notes that AIDS patients too have the tendency to develop some types of skin cancer.

So it is possible to say that, both above and below the micro-waves and radio frequencies, electromagnetic energy causes carcinogenesis. It would be a brave person therefore who argued that the radio frequencies themselves had no carcinogenic effects. The question is simply how strong do they have to be, and for how long does one need to be exposed?

In 1979 Bob Liburdy, who runs a research unit devoted to bioelectromagnetics at Lawrence Berkeley laboratory, testified to the harmful effects of radio frequencies when he found that radiofrequency (RF) radiation not only caused changes in T- and B-cell levels, but also reduced their immunocompetence (5.5:75). In a separate study of workers subjected to high voltages, it was found that fewer children were born to them than to controls, and that the difference increased with years of exposure (5.6:152).

Radar and Defence Communications

Concern about the effects of radar had already been expressed in 1964, by researchers at the Johns Hopkins School of Medicine, when they found that not only did children born with Down's Syndrome correlate with mothers given excess x-rays, but also with fathers who had been working near the microwave transmitters of radar stations (5.7:227). Radarmen were subsequently found to have higher than normal numbers of chromosome defects in their blood. In 1971, radar-exposed army helicopter pilots were also found to have been fathering chil-

dren with birth defects; within a 16-month period at the Fort
Rucken air base, Alabama, there had been 17 children born
with club foot–statistically there should have been only four
(5.8:126).

Radar beams (composed of pulsed microwaves) have the
highest power densities of any EMR source, states Bob Becker
in *The Body Electric*. Of more concern here is the possibility
that even very weak radio and microwave emissions can, over
time, cause health hazards not only to the present generation
but to succeeding inhabitants of our planet (5.8a:250).

Radar was in place round British coasts in 1938, just in time
to assist in the course of the aerial Battle of Britain. Young British
fighter pilots, though outnumbered ten to one, were able to know
the position of German bombers through the use of radar, even
though unable to see them. Soon afterwards, airborne radar
was developed by British research, and quickly exploited by
America, and this in turn gave way to over-the-horizon radar,
which is now being superseded by the new PAVE-PAWS systems, used to detect sea-launched ballistic missiles. This system is sufficiently powerful 'to resolve an object the size of a
football at a distance of over 1,200 miles' (5.9:20).

Unlike conventional radar, with its familiar rotating antenna,
PAVE-PAWS (Precision Acquisition of Vehicle Re-entry–Phased
Array Warning System) is equipped with an array of over 10,000
solid state radiating elements which can be individually
computer-controlled to form a single beam. This new kind of
radar was first installed in 1976 on the Cape Cod peninsula,
when its 500-feet-tall truncated pyramid of reinforced concrete
rose from the surrounding countryside.

As Paul Brodeur relates, the US Air Force had failed to sub-
mit an environmental-impact statement for the new Cape Cod
system, and when it did do so it adhered to the now obsolete
10 mW/cm permitted exposure limit (PEL) originally set by the
United States, instead of recognizing the Russian standard, pru-
dently accepted by the Environmental Protection Agency (EPA),
in its 1979 survey of radio-frequency exposure. Even then, PAVE-
PAWS could only comply with the United States standard by
time-averaging the measured power density emissions, and by
ignoring its peak pulse power values.

Finally, however, in May 1979, the Air Force were forced to
admit:

In view of the known sensitivity of the mammalian CNS [central nervous system] to electromagnetic fields, especially those modulated at brainwave frequencies, the possibility cannot be ruled out that exposure to PAVE-PAWS radiation may have some effects on exposed people (5.10:248).

Soon afterwards the Californian PAVE-PAWS system came into commission, radiating at 450 MHz, and rhythmically modulated at 16 Hz. During 1981 and 1982 Dan Lyle and Ross Adey were able to show that this frequency, when modulated, caused significant inhibition of T-lymphocytes' ability to identify foreign incursive toxins.

Meanwhile, the first AIDS cases were already becoming known, with skin cancers and with irreversible and declining numbers of T-cells–both symptoms associated with radiation. The first few AIDS sufferers lived on the American West Coast, where electromagnetic radio traffic is among the highest in the United States, and possibly the highest in the world. The next cases appeared in New York, which is said to consume as much electricity at any one time as the whole of Africa. As the cases built up thereafter, it became apparent that they correlated not with mere population, but with the density of electromagnetic traffic in the cities. Comparison of the latest CDC figures with the 1980 study of electromagnetic traffic in 15 major American cities by Richard Tell and Ed Mantiply graphically demonstrates this alarming correlation (5.11:243).

In Britain the first microwave towers were built in 1950, and according to Peter Laurie were part of air-defence preparations against attack by Russian aircraft carrying atom bombs (5.12:154). Cables are not capable of carrying the traffic effectively. From 1953 onwards the system was extended to guarantee the communications of central Government during an attack, and the GPO planned a chain of concrete towers code-named 'Backbone', to connect major cities by microwave link. In the sixties the system was again enhanced and the present system aims to serve the long-range warning stations at Saxa Vord in Shetland, Buchan near Peterhead, Boulmer near Alnwick in the Borders, Staxton Wold near Scarborough, Patrington near Hull, and Neatishead near Norwich. The route from Birmingham to Saxton Wold for instance, has a greater information-handling capability than that between London and Birmingham!

Defence communications now embrace satellite communications too. Typical of such systems is the huge 500-acre installation at Menwith Hill, North Yorkshire, code-named 'Steeplebush'. According to the ECP (Electronics and Computing for Peace) newsletter, it had 'at least 12 satellite communication dishes and other aerials and possibly up to 2,000 American service personnel, and is the largest known signals intelligence base in the world. D-notices have, however, kept it largely from public gaze. It has direct links with GCHQ (Cheltenham), Pine Gap Base (Australia), Forte Meade (Maryland, United States), and other main military centres. All voice and cable messages across the Atlantic are monitored by the CIA, as a result of SIGINT, a pact between the United States, Canada, Britain, Australia, and New Zealand. At Menwith Hill a complex of 18 DEC VAX11 computers look after this task, and these machines are part of a global network called Echelon. The 1989 United States defence budget stated that Menwith Hill station would have US $26 million spent on it over the next four years. In the United States this would call for an environmental-impact statement; in the United Kingdom no such statement has been issued to date. Similarly, elsewhere in the United Kingdom the expansion of microwave communications has been carried out unnoticed; at the Morwenstow base (run by GCHQ) near Bude in Cornwall for example, has been greatly expanded during the last ten years. In a letter to *Health Now*, Joan Rendell of Launceston wrote:

Mr Petley [a complainant] is not suffering from tinnitus; he has become one of the hundreds (probably thousands) of unfortunate people who are being driven slowly crazy by The Hum–courtesy of the Ministry of Defence. Gradually over the years (since March 1977) as the installations have increased and the signals have become stronger, it is within the hearing of a greater number of people.

Without going into long explanations, what Mr Petley is picking up is part of the early warning defence system, the noise being caused by transmitters and radar systems interacting with each other throughout the length and breadth of the United Kingdom.

As I sit here at 11 p.m. the hum is just building up. Like the Concorde sonic boom it is affected by atmospheric conditions and some nights is much louder than others...I and many others are gravely concerned over its effect on health, as being a microbeam (*sic*) it goes right through the head and there is no getting away from it indoors or out. ... There are certain areas of the country where Mr Petley can avoid it but not many.

Unfortunately in Cornwall we suffer horrendously with it
(5.13:212).

Launceston, where June Rendell lives, is only 20 miles from
the Morwenstow base, near Higher Sharpnose Point. Local wags
have renamed it Higher Sharpnoise Point.

In addition to the defence network and British Telecom's one
hundred and fifty or so towers, are similar networks operated
by the BBC, the Gas and Electricity Boards, Civil Air Traffic
Control, and the American Air Force. 'Often the towers of all
four stand within ten miles, and could just as well — and much
less expensively — have been a single tower,' says Laurie
(5.14:154). There are also two long-range military radio systems,
the first employing tropospheric scatter, and the second using
VLF (very low frequency) in the 10-16 KHz range. These need
huge aerials for their transmitters and there are three stations
in Britain, at Rugby, Criggion (Shropshire), and Anthorn (near
Carlisle). Rugby serves the Polaris missiles; Criggion is an older
system built during the war to provide an American link, and
may have been replaced by the complex web of aerials at
Woofferton, forty miles to the south; and Anthorn relays reports
from Fylingdales ballistic missile radar to the American air
defence headquarters in Colorado. These VLF systems are probably being replaced now by ELF systems which have greater
penetrative power. VLF can only penetrate nine metres into
water, whereas the ELF systems (which need even larger aerials, many kilometres long), can get much deeper, making sub-
marine communications possible. However, ELF systems
frequencies are in the vital band below 100 Hz and therefore
may interfere with the brain's own frequencies.

Satellite Communications

This chapter would not be complete without mentioning the
growth of satellite communications. In 1945, Arthur C. Clarke,
author of *2001 A Space Odyssey,* pointed out that stationary satellites orbiting the earth 22,300 miles above the equator could
beam television and other signals half way round the world.
This idea became reality in 1957 with the launch of the Russian Sputnik, and in 1962 the West's equivalent, Telstar, became
the world's first satellite to broadcast television programmes,
even though it was not actually in geostationary orbit. By early
1983 there were more than fifteen geostationary satellites for

the United States and Canada alone, and the total figure now exceeds one hundred.

The bands of frequency used by satellites are known as the C and K bands, the former being 3.7 to 4.3 GHz and the latter, which is used for Direct Broadcasting Systems (DBS), 11.7 to 12.7 GHz. Thus they are both effectively microwave systems. The essential difference between satellite and other microwave systems is that these new techniques involve cutting through the Heaviside layers of the ionosphere into space and back again. Since the lowest of the Heaviside layers is the ozone layer, the possibility exists that damage is being done to the ozone layer by such transits. As if to confirm the point, the areas of maximal ozone depletion seem to be above white surfaces of the globe (where radiation is most reflected) or above cities with high electromagnetic traffic. Even power line grids are known to have an effect on the troposphere.

Electromagnetic Communications and Health

Back on the ground, the energy received from these satellites is neglible, only about 8 watts of power at 4 GHz. But the energy of the uplink beams is high, and the side-lobes from the installations, can conceivably affect nearby populations. At the Royal Signals and Radar Establishment for example, the main radar beam is generated very close to housing estates nearby. At least three cases of streptococcal meningitis have been reported among infants within a few hundred metres, and there are several Down's Syndrome cases in the region.

The RSRE's South Site radar beam can easily be picked up at ground level; during my tests in nearby Britten Drive, a tidy little road adjoining the secondary modern school lined by two storey houses, it was being modulated at between two and three Hertz pulses–the same frequency as a sleeping infant's brain rhythms–so that I could not only detect it on my magnetic field monitor, but also on medium wave radio at about 500 KHz. When the radiating transmitter came to a halt and stopped revolving, I happened to have both devices operating, and it was quite surprising to find that I could also actually hear the change in my ears at the same time.

All radiation appears to be cumulative when absorbed by organic tissues–as if the tissues themselves are able to radiate the energy onwards to each other. The essence of all ionizing radiative elements is that they don't dissipate their energy all at once, but decay over a period of time. Some elements take longer than others. Their rate of decay is expressed in 'half lives'–the time it takes for half their energy to dissipate. As a result one shouldn't expose oneself to more than a certain quantity of x-rays in a year, because of these cumulative effects. One cannot help but wonder whether the same applies to non-ionizing radiations, since they are all part of one electromagnetic energy continuum, with no difference except in the amount of energy being expelled.

Following this line one might argue that chronic exposure to even very weak electromagnetic energy will gradually have the same effect as one short sharp burst. Even very low ('therapeutic') levels of ionizing radiation can have serious carcinogenic effects years or even decades later. Furthermore, blood which has been 'magnetized' with this energy will carry its magnetism with it by transfusion–as Jerry Philips discovered–for several cell generations, and proliferate accordingly until 'degaussed' (5.15:205).

Professor Herbert Frohlich, a physicist from Liverpool University, has researched the biological effects of microwaves for many years, and his technical papers on the problem are regarded as definitive (5.16:95).

Radiation in a frequency region that influences the coherent electric vibrations resulting in long range interaction [e.g. microwaves or millimetre waves] will interfere with this interaction [collective enzyme processes that oscillate with frequency] and hence with the subsequent processes including the EEG.

To the layman this simply means that microwaves jam the brain's own transmissions to its cells. He points out that the mechanism is already built in: 'The electromagnetic wave acts as a trigger to events *for which the biological system is already prepared.*' (My emphasis.) Frohlich's paper directly supports the notion of cerebral morphogenetic radiation (CMR).

Bob Dematteo, in his book on the health hazards of VDU screens, records:

Repeated low doses over a long period may present the same risk to health as a few equivalent high doses over a shorter period. However some of the evidence indicates that there may even be a supralinear dose-effect relationship. The effects are greater at chronically repeated lower doses than at short-term higher doses (5.17:73).

In 1948, when Israel became an independent state, one of the first acts of the Health Authority in an attempt to eradicate ring-worm from their new citizens, was to irradiate the heads of some 20,000 children suffering from this invading worm, *tinea capitis*. They administered very low doses (average 150 rads). Because each child had an individual population registry number it has been possible to follow up their progress twenty or more years later. When Elaine Ron and Baruch Modan completed this mammoth task, in 1988, they were dismayed to find that, even after twenty-six years, three times as many had died from tumours of the head and neck, and 2.3 times as many of leukaemias, as in the normal population. 'The central nervous system of the child', they concluded in a masterly understatement, 'is sensitive to the induction of cancers by radiation'. They left unsaid what level of radiation might be necessary for such induction, and how long it might take to produce its effect (5.18:217).

In another paper in 1988 Baruch Modan, who is the editor of a well-respected medical journal, was rather more pointed:

The relevance of such considerations goes beyond the nuclear industry to the full spectrum of energy resources. Nature hardly ever bestows a riskless benefit, and as in the case of ionizing radiation, it is becoming extremely difficult to isolate the benefit of EM energy from its potential risk. Consequently, even if the cost of precaution may eventually turn out to be unjustified, it highly outweighs ruthless self-confidence...EM energy, which is with us to stay, must be considered an environmental hazard until proven otherwise (5.19:182).

Words of wisdom indeed from someone who has spent a lifetime in radiation research, and who is one of the world's most respected radiation epidemiologists.

The world's exposure to RF or low intensity microwave energy over the thirty-five or so years since television broadcasting began has steadily increased in intensity. So much so that the British Astronomer Royal complained in 1988 that there were now so many radio stations taking up the wavebands that it was no longer possible to get a clear radio-astronomical picture of the sky. It is also almost impossible to be allocated a radio frequency today, since they are all already taken up by broadcasting stations. At the same time, the new generations of in-car telephones, CB radio sets, bleepers, local radio stations, and private telecommunications networks have so proliferated that we are now bathed in a sea of telecommunicating electromagnetic energy throughout the radio frequency spectrum. No wonder the pigeons, the warblers, and the whales are losing their way!

People living near a radio transmitter, microwave tower, or television mast should be aware of the risks they run from electropollution. A single case study will show the possible effects. Ludlow is a pleasant historic castle-town in Shropshire. It also happens to lie in line-of-sight of the complex towers of Woofferton a few miles to the south. In one house where I was asked by concerned parents to measure the effects of a nearby electricity substation on their health, I found that the entire rear of the house, where the bedrooms were, was irradiated by radio waves of substantial strength. The symptoms experienced by the family were typical; the previous elderly tenant had lived on the side of the house furthest from the signals, but she had become mentally ill, and was placed in institutional care. The present inhabitants' daughter was emotionally very volatile, and would cry at the slightest thing, sometimes without reason. Her father had become very depressed shortly after moving in, and this continued for some years until alleviated by transcendental meditation classes. The mother too suffered some emotional instability, and their son was troubled with learning difficulties. These relatively mild symptoms were evident within five years of their arrival, but had not been in evidence ever before. Though the electric field strengths in the family's bedrooms were not exceptionally high, at between 25 and 40 V/m, they were still about three times normal, and probably exacerbated by the eddy currents from the transformer substation, whose position directly between their house and the transmitters might have perturbed the waves even more. In time the effects would have become more severe, particularly as the imbalance of their ground return currents was noticeable from the high electric fields evident around their water taps.

My recommendations included moving the beds to avoid the highest fields in the bedrooms, exchanging their metal-framed beds for wooden ones, and changing to mattresses of natural fibres without any metal bedsprings; switching off the mains at night or installing a demand switch; and earthing the water system directly to the earth outside (a job which the plumber had forgotten). Other unused or inessential wiring in the bedroom was also removed, and their new knowledge of the house's fields helped them to find the safest places to sleep. It incidentally confirmed their own instinctive feelings that some rooms 'felt more comfortable than others to be in'.

In 1979 Richard Tell and Ed Mantiply of the Environmental Protection Agency completed a nationwide survey of population exposure to VHF and UHF broadcast radiation in the United States. They measured nearly five hundred locations round fifteen major United States cities, covering 20 per cent of the population. They found that 1 per cent of American citizens are potentially exposed to levels greater than $1 \mu W/cm^2$ which is the Soviet permitted exposure limit (PEL). The American PEL is a thousand times greater.

This means that some 2.5 million American citizens are being irradiated by television and related signals at levels which the Soviets consider too high for safety. In Britain and Europe, because of the larger populations and smaller geographical areas, the numbers similarly exposed are likely to be much higher.

The Russians should know what they're talking about. In the 1960s, by irradiating the United States embassy in Moscow with microwave energy of only up to $4mW/cm^2$ in power density (less than half the official United States permitted maximum even today), they are said to have caused that embassy to have the highest incidence of cancer in the world (5.20:20). Their research into the problem has been massive (5.28:213), and their results unequivocal; in June 1989 at a conference in Tucson, Arizona, they released details of a large-scale epidemiological survey which concluded that: 'It is imperative we find out more about the health problems encountered by people living in radio-frequency zones' (5.21:221) (5.21a:110).

There is no lack of monitory examples. At Forss, halfway between Dounreay and Thurso in Caithness, Scotland, the United States Navy operates a communications complex to send messages to aircraft, surface ships, and submarines, according to Louis Slesin, editor of *Microwave News* (5.22:229). There are a large number of transmitters at this base–some broadcasting in the highest frequency band (2-30 MHz), with one tall antenna sending signals at 50-150 kHz. At nearby Thurso a cluster of leukaemias in children has been reported near the Dounreay nuclear power station.

Six thousand miles away at McFarland, California, 13 children have developed cancer since 1975–way above what might normally have been expected. In Delano, up the road from McFarland, the Voice of America broadcasts at 9 and 11 MHz to Latin America and Asia. Its 250,000 watt transmitters, nearly ten times as powerful as the BBC's first Daventry transmitter, beam radio waves over the small McFarland community of 6,000 people. Outside another United States Navy communications centre at LuaLuaLei in Hawaii, between 1979 and 1985, children suffered incidence of cancer at four times the expected rate. In Gibraltar, where the British maintain a huge communications centre, the incidence of cancer among inhabitants is so high that a special recuperation centre has been set up in England to repatriate and convalesce the victims, who arrive there, it is said, at the rate of one hundred and fifty a year. In the small town of Vernon, New Jersey, there is a cluster of Down's Syndrome cases. Vernon is the site of one of the largest concentrations of satellite earth stations in North America.

In Poland, Dr Stanislaw Szmigielski of the Centre for Radiobiology and Radioprotection, Warsaw, revealed a clear association between non-ionizing electromagnetic radiation and cancer. He found that certain types of cancer related to leukaemia were nearly seven times more common among exposed soldiers. The younger the soldiers, the greater the observed effect. Only Poland has recently dared to monitor its military personnel in this way and report the results (5.23:241). Dr David Carpenter, of the New York School of Public Health in Albany, has estimated that 30 per cent of all childhood cancer may be due to power frequency (60 Hz) fields. From the above it seems possible that a fair proportion of the remainder may be due to irradiation of the public by microwave frequencies. The facts are beginning to seem sufficient to mount a legal action against those knowingly responsible.

It is quite possible that microwave frequencies induce severe depression; out of a list of 23 British microwave scientists and engineers who have died while working on defence-related projects since 1982, all except three, died in circumstances suggesting suicide. The coroner actually recorded suicide in seven of the cases (coroners tend to avoid recording this verdict, for the sake of dependents). The incidence of suicide among scientists at Marconi, Britain's largest defence contractor, and a leader in microwave communications, is said to be twice the national average.

The establishment response to the evidence of the harmful effects of microwave radiation is one of denial; Dr John Clark, one of Britain's most brilliant microwave scientists, died early in 1989 from cancer at the age of forty-four. He worked at the Royal Signals and Radar Establishment at Malvern, in an office directly below the rotating radar antenna. In the *Malvern Gazette* of 31 March 1989, Bob Lunn, a spokesman for the RSRE, was quoted as saying, 'There is no connection between the deaths from brain tumours of a number of scientists at the Malvern-based research station', and denied that microwave research had been halted at the base. A spokesman for the Ministry of Defence confirmed that no investigation was taking place, because the Ministry was satisfied with the stringent checks that were carried out at the base. Unfortunately, the official permitted exposure levels do not take account of proven damage to human cellular infrastructure, let alone emotional lability.

Dr Colin Watkins, of the airborne sensors division at the base, said at the inquest that 'regular checks were carried out every one or two weeks, and radiation was always considerably below the permissible limits'. Maybe the limits are wrong. Dr Clark lived less than two miles from the main radar transmitter at RSRE's South Site, near which at least three fatal cases of streptococcal meningitis were recorded. Sitting in my car in nearby Britten Drive, I could not only see the radar antenna revolving between the houses adjacent, but could also pick up its signals, modulated down to 3 Hz, as a powerful throb on my car radio. Curiously, I could even hear it physically, thereby involuntarily confirming some work carried out by Allen Frey in the sixties on the auditory response to microwave energy, modulated at low frequencies.

Tom Holland, Dr Clark's predecessor, also died of a brain tumour, as did Tom Dunmore, who worked in the department, which consequently has one of the highest incidence of brain tumours per head of population in the Western world.

AIDS

Before discussing what one can do about it, it is important to recognize the possibility that electropollution may be an important causative factor behind our most serious immune deficit, AIDS. I have not space here to set out in full the argument for an electromagnetic aetiology of AIDS, but I can say that others also are questioning the manufactured opinion that AIDS is viral in origin. Molecular biologists of formidable and world-wide reputation, like Peter Duesberg, point out that the facts simply do not support a viral explanation, and that there must be other co-factors at work (5.24:81).

The same opinion was expressed as early as 1985 by Richard Ablin of New York State University (5.25:2). In 1989 there were several letters in the *Lancet* recording cases of serious immune deficit without any sign of HIV, and numerous other cases are on record of people with HIV but no sign of AIDS (5.26:71). One telling experiment found recently that HIV was proliferated 150-fold by ultraviolet radiation, and 50-fold by sunlight alone, pointing to an electromagnetic connection (5.27:252). It is unnerving to discover–having already discussed the possibility that neonatal electromagnetic irradiation might cause cot death–that two thirds of American AIDS cases were born in the late forties and early fifties, the very years when microwave telephone systems were being installed by Bell Telephones across the nation (5.28:43;52). In Britain the same pattern occured a few years later (5.29:9). Even an early case of AIDS in 1968–only diagnosed subsequently when the New Orleans-born 16-year-old victim's blood was examined decades later–was, like the others, born in 1952 (5.30:268).

Remarkably, no research has yet been carried out into the possibility that AIDS is acquired neonatally as a result of electromagnetic insult, despite the growing evidence that almost every other kind of immune deficit can be ultimately related in some way to electromagnetic radiation, either ionizing or non-ionizing. The threat to humanity posed by AIDS is so immense that no stone should be left unturned in attempts to find a cure. If AIDS continues to double every fifteen months, as is the case at the moment, then in less than ten years some 50 million people will have succumbed worldwide. A million of those cases will be in Britain and you will in all probability personally know someone who has died or is dying from AIDS–if it is not yourself, of course.

When Jad Adams published a book in 1989 questioning what he called the HIV myth (5.31:5), his arguments were broadly accepted, with a few objectors in medical journals. The substance of the objections however, when examined in detail, was confined to minor matters, and the main thrust of his argument was unchallenged (5.32:206). The fact is that AIDS results from a disappearance of T-cells, and HIV may not be the cause of why those cells disappear, since injecting someone with HIV does not produce AIDS in that person. This is one of the prime tests for any viral aetiology; it was postulated by Robert Koch a century ago, and HIV does not satisfy it. Indeed, after the State of Massachusetts had screened all new babies for HIV, they found that 1-2 per cent of them were HIV positive, irrespective of which social group or geographic location they came from. Moreoever there are many varieties of HIV (or HTLV–human t-lymphotropic virus–as it was previously and more accurately called). Which one of them is responsible? The first two types have already been associated with leukaemia (5.33:70), and leukaemia in turn with electromagnetic fields. Why not the third type of HLTV? As Connor and Kingman say in their book *In Search of the Virus*, no one knows where AIDS came from, yet there is evidence that it started simultaneously in many parts of the world (5.34:60).

Though it might seem strange to find a report on low level radiative emissions from the Sellafield nuclear reprocessing plant discussing the AIDS virus, the connection between radiation and AIDS is never far from the surface. Sir Douglas Black's 1984 report on cancer in West Cumbria says:

Human adult T-cell Leukaemia virus (HTLV-1 or ATLV) is the only human virus that has been implicated in human leukaemia to date. This is a retrovirus which induces an aggressive variant of mature T-cell leukaemia. HTLV-1 has so far been identified as occurring mainly in South West Japan, in the Caribbean basin, in Central America, and in Africa (Gallo 1984). [He does not notice that all these areas of the earth are where solar radiation is at its highest.] The natural mode of transmission of these agents has not yet been established. More recently a new virus, LAV or HTLV-3, has been identified as the probable cause of Acquired Immune Deficiency Syndrome (AIDS) (Weiss, 1984). HTLV-1 and HTLV-2 have not been detected in any cases of childhood leukaemia (5.35:30).

His last sentence helps the argument that the immune deficit was in place before the virus arrived.

Leukaemia

I mentioned in Chapter 4 that the types of leukaemia cases living near the Sellafield lines were generally lymphocytic, or

lymphoblastic–lymphoblasts are immature lymphocytes–
whereas those near to radio and television masts or on hills
seemed to tend towards the myeloid type. Louis Slesin points
out that the Thurso cluster of leukaemias are all in the same
housing estate, situated on a hill where EM radiation would
be greater: 'Since radio transmitters aim their signals in the air,
the power density of non-ionizing electromagnetic radiation
from them is likely to be greater on higher rather than lower
ground,' he says. I have observed quite a few cases of ME on
exposed ground near to radio masts.

This observation deserves enlargement. Leukaemia does not
always happen swiftly; sometimes it is preceded by what is
known as the preleukaemic syndrome, whose symptoms are
non-specific weakness, lassitude, weight loss, and sometimes
a tendency to bruise easily. (These symptoms are similar to
anaemia, and also to ME–another, milder, immune disorder.)
Leukaemia is a neoplastic disease of the bloodforming cells,
with a large variety of types. Chronic forms of leukaemia have
a duration of several years and involve a proliferation of mature
cells, whereas the acute forms only last a few months and impli-
cate the blood cells before they have reached maturity. In both
kinds there is a progressive infiltration of normal bone marrow and other body tissues by leukaemic cells.

Why should this happen? The only aetiology ever proven for
leukaemia in human beings is that of radiation. To quote an
eminent haematology textbook:

Radiation exposure is a predisposing factor to leukaemia. This
was established by the studies of survivors of atomic bomb
explosions in Japan where an increased frequency of acute
lymphoblastic as well as acute myeloid leukaemias occurred.
Diagnostic x-ray procedures in the mother at the stage of the
early embryo doubles the chance of leukaemia in the offspring
(5.36:44).

When it is realized that these ionizing radiations themselves
give off ordinary electromagnetic energy, then the connection
between leukaemia and non-ionizing EM energy becomes
nearer to being proven.

Ionizing radiation consists of:

(i) alpha particles: these are helium atoms without their two
outer electrons. They therefore carry a net positive charge.

(ii) beta particles: these are high energy electrons and are therefore negatively charged. If stopped suddenly their kinetic energy is converted into heat and electromagnetic radiation (x-rays).
(iii) gamma particles: these are a form of electromagnetic radiation with a wavelength shorter than that of visible light. They are similar to x-rays except that x-rays are produced by collisions of high-speed electrons while gamma rays are simply emitted from radionuclides.

About 90 per cent of acute childhood leukaemias are lymphoblastic, and only 10 per cent myeloblastic. In contrast, chronic lymphocytic leukaemia occurs in the last few decades of life, 90 per cent of patients being over fifty.

   Why are the human beings affected by lymphocytic leukaemias making so many more lymphocytes than normal? First it should be said that of these excess lymphocytes, 80 per cent are 'null-cells', 20 per cent T-cells, and only 2-3 per cent B–cells. The null-cells, only found in cases of lymphoblastic leukaemia, seem to have 'dispossessed' of their function. It was found in the 1960s by Madeleine Barnothy that if rats were exposed to magnetic fields, their immune systems, as measured by a leucocyte count, were at first depressed by up to 40 per cent, but then a surfeit of lymphocytes magically appeared as if in compensation, and it took up to three months to get back to normal levels:

The first experiments were made in the laboratory of the Biomagnetics Research foundation in 1956. They revealed that a homogeneous vertical magnetic field of 4,200 Oersted does indeed increase the number of circulating leucocytes of the mouse in the first week of residence in the field, but does not observably affect the erythrocyte count. After removal from the field a recovery sets in, during which the leucocyte count overshoots the base line. Several further experiments performed since then have confirmed the previous results (5.37:16).

Unlike human beings, the leucocytes of a mouse are two-thirds composed of lymphocytes, and since Barnothy was using a Coulter counter, which counts the cells electronically, her results are unquestionably accurate. She suggests that the reason for the excess is a temporary mobilization of the stored cells in the marrow, which are however, still immature, and they too succumb to the impact of the field.

The myeloid proliferative group of leukaemias is also characterized by a surfeit of white blood cells, and they are related
in the sense that polycytaemia may transform into chronic
myeloid leukaemia or evolve into chronic myelofibrosis. These
are disorders resulting from mutation of a single stem cell, and
all are marked by a sharp increase in granulocytes (white cells
with many granules round the nucleus).

We are seeing two different mechanisms at work here: in the
lymphocytic group the lymphocytes' ability to discern foreign
cells has been impaired through radiative 'jamming', and there
has been a consequent accumulation of immunologically
incompetent lymphocytes, whereas in the myeloid group the
radiation has actually caused a mutation which is then multiplied by mitosis as the mutant cells divide and proliferate. It
has been shown that mutations and mitosis generally are frequently dependent (5.38:61), so it would make sense to find
one kind of leukaemia clustered round power lines and a different kind near to radio frequency sources.

I was curious about this possibility, for I remembered how,
in another study, non-lymphocytic leukaemias were not found
to be associated with power line proximity (5.39:222). My search
through the literature uncovered another piece of evidence to
support this hypothesis; James Lin, of Detroit's Wayne State
University Department of Electrical Engineering, had been carrying out research into the effects of exposing granulocyte
precursor cells from mice (which usually form colonies by mitosis) to microwave radiation. With increasing microwave exposure
he and his team found that mitosis (or cell division) in the
granulocytes was curtailed. 'It is interesting to note', he reported,
'that the same exposure system and parameters had no effect
on the growth and viability of fibroblast and lymphoblast cul-
tures', and referred to a similar result which had been obtained
by a Polish research team (5.40:158).

If these scientists are right, microwaves will affect the granulocytes and hence lead to myeloid leukaemias, whereas radiation
at power frequencies is more likely to affect lymphocytes. This
provides us with a clue to the frequency dependence of these
two kinds of white blood cells, suggesting that their mitosis is
controlled by two quite different frequencies, well apart on the
electromagnetic spectrum.

This means that while living near power lines might induce
acute lymphocytic leukaemia (from which the chances of recovery are more than even), living near radio or microwave radia-

tive sources might induce myeloid leukaemia, the prognosis for which is much less optimistic. It would also follow that the lymphocytic group of leukaemias can be reversed simply by evading the field for long enough, whereas the chances of recovery from myeloid leukaemia must rely on correcting the mutated cell lines. This is in fact the case; an abnormal chromosome, the Philadelphia chromosome, is present in the blood of 92 per cent of patients with chronic myelocytic leukaemia, and it persists. By contrast, in lymphocytosis there is infrequent cell division, since the disease involves an increasing accumulation of long-lived but ineffective lymphocytes, which is how leukaemia (literally 'white blood') gets its name.

In this regard, Conti's, and Lyle and Ayotte's, separate discovery that mitosis of human lymphocytes is markedly inhibited by exposure to EM fields is significant. The incidence of acute lymphoblastic leukaemia (the commonest leukaemia of childhood) has increased in incidence in the last thirty years, during which time the use of domestic electricity has also increased many times. (The emergence of antibiotics will have mitigated this trend, since intercurrent infections have doubtless been reduced which otherwise would have proved fatal before the leukaemia became apparent.) As a final piece of supporting evidence for the argument that leukaemia is caused by electromagnetic energy at non-ionizing frequencies, acute lymphocytic leukaemia is seen occasionally in adults, particularly those working in electrical environments (5.41:156), whereas myeloid leukaemias are noted most frequently among telecommunications workers (5.42:180).

In tropical countries, and also at high altitudes (both places where the effects of solar electromagnetic energy are more pronounced), a malignant disease of lymphoid tissue known as Burkett's lymphoma occurs, but is found nowhere else. This disease is remarkably similar to acute lymphoblastic leukaemia.

The symptoms of myeloid leukaemias and AIDS are similar; sweating at night, tiredness, loss of weight, etc. Evidence of retroviral existence is also found both in AIDS patients and in 95 per cent of all types of human leukaemia, which persists long after remission has taken place. Yet no one has ever tracked down any virus associated with human leukaemia, or been able to show that it is transfective. I believe that there is an electromagnetic or radiative aetiolgy behind both these modern disorders of the immune system, and logic supports my view.

The traditional treatment for lymphocytic leukaemia is Vincristine, or similar drugs, which actually kills the excess lymphocytes without impairing marrow function, followed by irradiation of areas where the Vincristine cannot reach, like the skull, for instance. Remission is thereby achieved in 80 per cent and an apparent cure in over 50 per cent of cases.

My suggestion is that the patient should also be completely shielded from any form of electromagnetic radiation. The lymphocytes then have a chance to normalize, and since they are generally short-lived cells, a recovery might be seen within a few weeks. More importantly, the source of the radiation must be identified and either removed (e.g. switch off the mains electricity at night), or avoided (e.g. don't sleep near an immersion heater or on an electric blanket). This is only applicable to the lymphocytic group.

The myeloid group requires a twofold approach; first to see whether there are any obvious radio or microwave frequency generators nearby, and if so establish thick Faradic protections against their electric fields, at least. To protect against magnetic fields is extremely difficult; it is better to give up, and simply remove the body from the offending field. Secondly, the mutant cell strains which are proliferating inside the victim's body must be eliminated. Most mutant strains of cell are normally disposed of by the immune system. But here we are dealing with an immune system which is itself faulty. Moreover, it is normally a disorder of elderly rather than young people, whose cerebral morphogenetic radiation systems are no longer at their best.

The only technique I know of here which might prove effective is known as the Knott technique (5.43:114). This was used to great effect before the advent of antibiotics made it somewhat cumbersome and obsolete, but I note that it is enjoying a revival in Germany these days. The patient's blood is irradiated with ultraviolet light and then replaced. Only about one–twelfth of the blood volume is actually treated this way, but since the UV is not only bacteriocidal but also can kill other cells of the blood it presumably has the effect of damaging the mutant cells, or perhaps breaking the aberrant hydrogen bonds which are causing the mutation, after which the brain can better correct the DNA in the cells with its own unique normalizing propagations. Furthermore, the irradiated blood secondarily irradiates the blood which has not been withdrawn from the body. The Knott technique of haemo-irradiation was invented as early as 1934. It has been used to treat septicaemias, atypical pneumonias, herpes-related diseases, poliomyelitis, osteomyelitis, and many other diseases opportunistically profiting from a weakened immune system (5.44:177) (5.45:178).

It would not surprise me, once the electromagnetic aetiology of AIDS has been accepted by the medical profession, if the same technique could be used to advantage in the curative treatment of AIDS itself.

Meanwhile, to avoid undue exposure to radio frequency irradiation, seek out bedrooms with strong thick walls! A late-nineteenth century explorer was once accorded the rare honour of sleeping in the King's Chamber of the great pyramid of Gizeh. This chamber is not only buried deep in the centre of the building, and aligned carefully along the north south axis of the earth's magnetic field, but is roofed by no less than six enormous silicate granite blocks, as if purpose-built as the world's earliest electromagnetic 'clean room'. He reported the next morning that he awoke 'wondrously refreshed' (5.46:155). No doubt he had had the benefit of a full overhaul of his cells during the night, when a more than usual amount of protein synthesis and mitotic repair had taken place.

I close this chapter with the suggestion, albeit a bizarre notion, that the now disused coalmines of South Wales may be brought back into service to assist in the repair of immune systems weakened by today's electromagnetic environment. Only some hundreds of feet below the surface is one assured of escaping the possibly hazardous influence of radio and microwave irradiation; unless like Dr Boyd, one clads one's entire bedroom in copper sheeting!

6. Sick Buildings and Healthy Houses

Great cities will be deserted, and not one soul will live in them.

Nostradamus, Century Three, Quatrain Eighty Four, 1611 A.D.

An architect, Michael Schimmelschmidt, and I once organized a conference at London's Hale Clinic with the title which heads this chapter. In it Michael, borrowing from Dr Karl Ernst Lotz, discussed the idea of the building as human beings' third skin, the second being the clothes which, through evolution, humans have chosen to wear:

A building is more complex than our clothes, and has to take care of our other two skins. It is a holistic approach to the interaction between life forms and the living environment. We can contribute to the well-being of the body by carefully choosing the location of a building, avoiding geopathic stress zones, protecting its inhabitants against electromagnetic fields, paying attention to the geometry, general arrangement and shape of the building, selecting natural non-toxic and ecologically-sound building materials. Colours, texture, smell, the heating system, all can have a very positive influence on our organism (6.1:161).

In this chapter I will deal with the aspects of the building which concern the flow of artificially-created energy through it.

Buildings and Air Ions

Any building is a superstructure above the earth's surface. As a result it is subject to certain physical phenomena, including the observed fact that ions (electrons freed from their atoms, and travelling about in search of another home) tend to congregate at the sharper points of the materials in which they are resident. An example of this is seen in the apparatus called the van de Graaff generator. In this instrument ions created by friction congregate at the surface of a sphere and their charge builds up until the intermediate space between the sphere and another body breaks down, causing a large discharge spark. There is a van de Graaff generator in the London Science Museum, which children can operate, by turning a handle until the spark occurs.

The same phenomenon is seen in pyramidal structures; if you put a metallic razor blade inside such a construction the ions congregate along the razor's edge, and people claim that it is thereby re-sharpened. Above a pyramid, the presence of ions causes eddy currents; American pilots flying over Egyptian pyramids claimed that their instruments went haywire.

Tall buildings are subject to the same ion flow, and since the earth's surface is mildly negatively charged there should be slightly more negative ions at building tops. Mountains also contain more negative ions, and their beneficial effects make them a holiday favourite, second only to the seaside, where negative ions are also generated from the tumult of wave upon shore. Instinctively, we seem to know that such places are good for us.

When a storm front comes along, by contrast, it pushes ahead of it a positive ion cloud. Accordingly people begin complaining that the atmosphere is unpleasant or 'close'. After the storm is over, and lightning (Nature's own van de Graaff generator) has discharged the polarity built up between the negative earth and the positively-charged underside of the storm clouds, a surfeit of negative ions is left behind, which is why air seems so clear after a storm.

Electrical Machines

Negative ions are good for us, and there is evidence that positive ions are not (6.2:233). Unfortunately, in cities the number of negative ions is markedly reduced, and particularly so in offices where the electrical machines are continually discharging positive ions. Accordingly, office workers are at risk of working in a unhealthy atmosphere, unless the situation is remedied.

I was recently invited to sit in on a BBC studio and watch the preparation of a broadcast. The room was windowless, the overhead lights were screened with metal, and the ventilation was served by a ducted chute whose air would almost undoubtedly have lost enough electrons en route from the outside to be a source of many positive ions. Other positive-ion sources were the complicated banks of tape decks, amplifiers, control consoles, and speakers, constituting a whole variety of electrical and electronic equipment.

I was not surprised to see the tape-deck sound engineer constantly massaging his lower back, the producer's assistant trying to relieve the tension in her neck, and the studio manager's face a battleground of spots and skin eruptions. The presenter, safely incarcerated next door in his glass-windowed soundproofed room, who also spent much of his time outside on location, displayed none of these symptoms. I have no doubt that a suitable negative ion generator in that studio would have made life a good deal more bearable for its inmates.

Ions and Health

A controlled study in 1981 by Leslie Hawkins of Surrey University's Department of Occupational Health found that when a negative ion generator, or ionizer, was surreptitiously installed in offices the incidence of headaches, nausea, and drowsiness fell by 50 per cent. Nightshift workers (nowadays all-night programmes are a common feature of radio), received particular benefit from the negative ion generator (6.3:121).

Ions can be created from any gases in the air, but most seem to be charged forms of oxygen or water. In the United States, the FDA has curiously banned ionizers for medical purposes since the mid-fifties, partly because early manufacturers made extravagant claims. One solitary American researcher, Igho Kornblueh, who continued his work on air ions despite considerable opposition from vested interests, claims that ions affect the electroencephalogram and are of clinical value (6.4:148).

Sulman (6.5:237) (6.5a:238) in Israel and Krueger (6.6:149) in the United States have revived interest in air ions, and Fred Soyka's book *The Ion Effect*, which describes in layperson's terms their work and the generally beneficial effects of negative ions, has become a bestseller (6.7:233). Even so, the handful of United Kingdom manufacturers today have only penetrated 7 per cent of the potential market, so great is public ignorance about ions and their effects.

Our blood cells by contrast, know only too well the benefits of negative ions! At the time of the full moon, it was observed by Dr Norman Shealy, post-operative bleeding is most severe.

At this time there are always more positive ions in the atmosphere, because the moon, pulled nearer the earth by the sun's gravitational force, causes the ionosphere to be somewhat squeezed towards the earth, and the underside of the ionosphere is positively charged (see also Best and Kollerstrom's *Planting by the Moon* (6.8:27). Soyka tells the story of an enforced experiment on the effect of air ions on buildings. In the town of Overgard, a farmer, Flemming Juncker, had two monstrous chicken houses, each holding 20,000 hapless birds. In one of them, some 200 birds died each week, whereas in the other mortality was minimal. The roof of the affected chicken house was lined with plastic, but the other roof was wooden. Whenever there was a change in the weather, reports Soyka, mortality increased. Suspecting that the resulting increase in air ions might be to blame, the researcher, Christian Bach, treated the plastic roof with an antistatic coating, and within weeks the mortality in both houses had fallen to the same low level.

All around the Western world are modern buildings which create problems for their inhabitants, known as sick buildings. The new Automobile Association offices at Basingstoke are said to be sick, as are the new Council offices at Angers and the modern offices at some time occupied by Rothschild in Paris.

Symptoms include tension, lack of energy, depression, and headaches. Ion counts in cities are almost always low in negative ions, since these are depleted by atmospheric pollution and airborne particles. The ductwork in buildings which brings in air from outside should not contain sharp or right-angled bends, since these cause the friction which knocks off the electrons, leaving positively charged atoms to continue into the office atmosphere.

Hospitals can be affected by the plethora of electronic monitoring equipment now used in operating theatres and observation rooms. When I measured the ambient electric field density in the Special Care Baby Unit at the Royal Gwent Hospital in Newport, South Wales, I found that the neonatal incubators (whose environment is carefully controlled for the preservation of premature newborn babies), were in a permanent electric field of over 30 Volts per metre–some three times the norm–simply through their proximity to the sophisticated monitoring equipment there. Moving them even one metre further away from the equipment would have reduced the exposure, and probably improved survival rates.

Incidentally, I am told that survival rates improve if the mother is actually allowed to hold the baby for a while, or given the opportunity of close contact. Her own special electromagnetic signal acts as a shield for her offspring. Some supporting evidence for this notion is offered by the common observation that a sleeping baby will synchronize its breathing with that of its mother, and since there is no physical connection, there must be some sort of communicative action at a distance. Women also find that when they live under one roof their menstrual cycles gradually synchronize. The cerebral morphogenetic radiation can satisfactorily explain this phenomenon, on the premise that their inter-cerebral communication is electromagnetic in nature.

All these mechanisms are little researched, but point, like telepathy, towards an intercommunicative mechanism between members of one species. Good genetic ideas may be intercommunicated in the same way; Darwinism does not by itself explain the rapid speed of evolution through survival of the fittest and natural selection. Something is lacking, as Rupert Sheldrake has pointed out (6.9:224).

People who want children, yet do not seem able to conceive, often do so on moving house; 'New house, new baby' is a tried and trusted adage. I was asked once to design a protective garment for would-be mothers which would protect them from electric-field insult during the vital early stages of mitosis after the fertilized ovum had been replaced surgically into their body. I gather that afterwards the success rate, previously only 13 per cent, improved dramatically.

Hartwin Busch, who has spent many years designing ecologically-fulfilling houses, sees the building as a skin, like Michael Schimmelschmidt. But he points out that although our skins protect us from ultraviolet radiation, they are also permeable, letting out perspiration, and letting other substances, like oils and embrocations, in. It is said that an aromatherapy oil can be detected in the urine a few hours after application (Worwood, 6.10:269). Busch explains that many of our modern sick buildings have impermeable walls which cannot breathe. (Nor presumably can they admit negative ions from the atmosphere, as a result of using new materials like plastic and glass and concrete, instead of porous materials like brick and stone.)

The old French chateaux were so designed that their walls absorbed a certain amount of moisture, which then evaporated and kept the structure cool in summer and warm in winter, like a porous clay butter-dish. Any attempt to build a damp proof course into them or seal the exterior caused a sudden increase in decay and a less pleasant interior atmosphere.

Modern synthetic materials like nylon are well known for their susceptibility to electrostatic charges–as anyone walking across a nylon carpet towards a metal door handle is about to find out! These same charges are also impacting on the body's cells, and though small, are not conducive to well-being. Natural fibres like wool by contrast, are themselves composed of cells, and are thereby more easily electrically neutral; the electric barrier across an organic cell membrane is formidable, in the region of some 20,000 volts per metre. People have soon found out for themselves the superiority of natural cellular fibres; the cost of wool and other natural fibres is reflected in their higher retail price. The synthetics industry, for all its growth, has not been a complete success, for today an acrylic sweater or nylon carpet will sell for only half the price of its woollen equivalent. Sadly the natural fibres are being edged out; linen shirts, so much cooler than cotton on a hot day, are prohibitively expensive and hardly ever stocked. In the eighteenth century a linen nightdress was a common comfort for the humblest shepherd. Plastic has replaced wood in many building applications. The modern obsession for warmth has lengthened air recycling periods, inevitably worsening the positive ion ratio in rooms.

Electropollution in Buildings

Electropollution in buildings is every bit as subtle as ion quality, and indeed results from the escape of ions into the atmosphere by induction from the current-carrying wires. Every domestic wiring circuit induces an electric and magnetic field of ions round it. Normally the electric kettle or toaster we switch on only generates its field for a minute or two but the alternating current in the ring main supplying these appliances is there day and night.

Their insidious damage is at its most pernicious at night; it is during this time that the brain is instructing for protein synthesis in order to repair the 500 million cells lost in the day by mitosis. I suspect that only the brain's CMR signal can initiate the mitotic stage of the cell cycle, and low intensity fields from domestic wiring disrupt these signals. Amazingly, if a light bulb is switched on in some part of a building, the EEG records of human beings in that building register the change.

Fluorescent Lighting

Accordingly there are a number of scientific studies which correlate lighting of certain types, noticeably fluorescent tubes, not
only with carcinogenic effects, but also with other health hazards because they contained PCBs (polychlorinated biphenyls)
(6.11:23) (6.12:120) (6.12a:76) (6.12b:87). The latter is outside
the scope of this book, but some indication of their importance
is given by the story that in 1983 vigilant staff at two Berkshire
schools noticed small oily pools on the floor below fluorescent
fittings. The pools were PCBs which had leaked out from the
classroom's fluorescent fittings. The local authority actually shut
the schools as a precautionary measure during the clean-up
operations. The public analysts at Reading noted at the time
that they were:

A potentially serious hazard in the buildings carrying the
fittings. All these substances are very toxic, and there could be
damage to health in the event of leakage. The possibility of quite
large claims against the county council cannot be altogether
discounted (6.13:139).

The even greater, carcinogenic, hazard from induced electric
and magnetic fields emanating from such tubes, by contrast,
has not been given the same urgent consideration. Even the
London Hazards Centre booklet on fluorescent lighting says,
for example; 'The low energy levels from fluorescent lights are
unlikely to cause thermal effects, but the nervous and reproductive systems may still be affected,' and leaves it to the reader
to consult a reference in a footnote (6.13a:144). The symptoms
listed by the LHC, however, do include an increased likelihood
of cancer; they cite at least nine scientific papers on the subject. Of these, perhaps the most important are the Canadian
Centre for Occupational Health and Safety's abstracts, published in 1981. Valerie Beral and her team produced a startling
paper on the relation of malignant melanoma and exposure to
fluorescent lighting at work as early as 1972; this was published
in the *Lancet* (6.14:23), and thereafter largely ignored, despite
Beral's pre-eminence in epidemiological circles. The wider
implications of these human chicken houses are discussed in
Chapter 8.

Fluorescent tubes emit the same sort of radiation as VDUs,
indeed their method of manufacture includes a glass tube internally coated with phosphorus just like a computer or televi-

sion screen, which in the case of the lighting is activated by an electrode at each end. The phosphorus converts electromagnetic energy into visible light. It also, it seems, makes organic cells up to a thousand times more sensitive to RF radiation, according to some unpublished work done at London University in the sixties by scientists under Professor Charles Turner.

Unbalanced Electric Fields

Apart from fluorescent lighting, there are dangers when the electric currents are unbalanced in some way. Electric fields, no less than magnetic fields, are more or less balanced in alternating currents when the ground (neutral) return wire is adjacent to the 'live' wire. The problem arises when they follow separate paths. As Brian Maddock and John Male, of the erstwhile CERL (now National Grid Research and Development Centre), explained in a recent paper:

In homes, fields of up to a few tens of microTeslas are detected close to appliances, but they fall away rapidly with distance and beyond one or two metres have usually become less than the typical United Kingdom domestic background of a few tens of nanoTeslas. This background field, which usually fluctuates markedly with time, seems to arise mainly from distribution cables under streets. Such cables normally comprise three helically-laid conductors for the phases and a neutral conductor or a neutral sheath. In a simple case current in the neutral would be equal but opposite to the vector sum of the phase currents so that the external magnetic field would be negligible. However, the modern practice of earthing the neutral at more than one point means that some current may flow in the ground or other return paths so that the cables are not quite balanced. A few amperes out of balance is sufficient to produce the background magnetic fields observed (6.15:164).

Wertheimer and Leeper found that the current delivered to a house often returned to the mains via the ground rather than through the wire designed for the purpose. These unbalanced ground returns can be found in a variety of situations, even within the walls of a house, and can produce some surprisingly high electric fields, especially where metallic materials have concentrated the fields. Aluminium backed plasterboard, ribbed steel joints, and even radiator systems can all cause abnormally high ambient fields.

In multi-storey blocks, for instance, the rising cable might ascend on one side of the building and return via the various circuits of each apartment some distance away. In 1987, Dr Stephen Perry, a physician based in Birmingham, took a six-month sabbatical to carry out a study of these effects in multi-storey blocks. His co-worker was Laurence Pearl, a statistician from Wolverhampton Polytechnic, and they used data from 1985-86 relating to 49 high rise blocks with over 3,000 housing units. A total of 6,000 occupants lived in the blocks, 37 of which had electricity supplied via a single rising cable at each end of the building. In only 6 of the total blocks was electricity supplied individually to apartments.

Perry and Pearl divided the occupants into two groups, those living near and those living furthest away from the rising cable, and not surprisingly confirmed that the 'near' group were on average exposed to significantly higher electric fields than the other group. It had been the expressed opinion of the CEGB that the fields were highest on the lower floors, but this was not found. The results of this study were compelling; the group living nearest the rising cables suffered a much higher incidence of depression than the others (22 cases out of 31). Where electric underfloor heating or storage heating systems were in use, they noted, the incidence of depression rose to 82 per cent (6.16:201), which supported an earlier study by Perry where elevated levels of suicide had been found among people living near high-voltage cables (6.17:200).

The term 'building biology' ('baubiologie' in German) was coined by Dr H. Palm of Konstanz in 1955. His book *Das Gesunde Haus* (The Healthy House) has since seen nine editions. Even then he was advocating that the mains electricity should be turned off at night, at least in the sleeping area, and that babies should not be put down to sleep near television sets (which accumulate voltages up to 15,000 volts unless their plug is removed). Dr Palm points out that our house is our closest environment, and twenty times as important as the external environment about which we voice so much concern. At the beginning of this century workers spent some 70 per cent of their time out of doors, but now, thanks to electric light and other technology, we spend less than 10 per cent outside (6.18:197).

The use of a demand switch or shielded cabling will quickly ensure that current only alternates in our bedrooms when we are actually consuming current, or that the fields it propagates are too low to cause effects. Another protection is to shield the cables themselves in a conduit which should itself be earthed, or to use braided cabling so the return currents are fully balanced. The disadvantage of simply turning off the mains, apart from being a nightly chore, is that it might cause food to spoil in the freezer, and it makes it necessary to keep a torch by the bedside, as well as to forgo our radio-alarm clock with its digital panel. The video too has a clock which will blink at us in silent disapproval until reset. However, a little ingenuity can overcome these minor disadvantages for the sake of our long-term health; battery-driven alarm clocks and radios are available, plastic bottles of water in the freezer when frozen will maintain a low enough temperature for hours, and separate circuits well away from the bedroom should not be harmful.

Natural Radiation and ME

The work of Dr Karl Ernst Lotz on building biology is summarized in his book, *Do You Want To Live Healthily?* (6.19:161). He points out that natural radiation from the soil can affect buildings in noticeable ways. While animals seem to sense and avoid noxious radiation, plants have no means of controlling where their seeds fall, and their subsequent growth can provide a natural barometer of radiative hazard. Trees growing over subterranean aquifers show cancerous growths, claims Lotz. An unnatural shooting up of branches–so that a cherry tree, for example, looks more like a poplar–can indicate a similar hazard; or simply a persistent withering of planted vegetation, the 'blasted heath' deliberately chosen by the witches in *Macbeth* for their rendezvous may be a famous example (6.20:14). 'The earth has bubbles as the water has', says Banquo prophetically, as the witches dematerialize.

The human body's galvanic skin response (GSR) automatically changes in response to an artificially-applied current, even of imperceptible intensity. The variation between standing over an underground stream and on a bridge above open water can be between 13 and 23 kiloOhms. Since the current produced by the underground movement of water forms a continually changing magnetic field around itself, it constitutes a chronic disturbance of the environment.

I have found that important clusters of myalgic encephalomyelitis nearly always occur first in dormitory residences near such subterranean aquifers (6.20a:99;228). For example the famous 1955 outbreak at the residential school for trainee nurses at the Royal Free Hospital, London, which gave the disorder its first name, was only a few metres from Fleet Road, a steepish hill down which runs the now conduited Fleet river on its way from Hampstead Ponds to the Thames (6.21:11). Similarly the 1970 outbreak at the Childrens' Hospital, Great Ormond Street, occurred first in the nurses' dormitories adjacent to Lamb's Conduit Street, a few short metres from the underground route of an old but still existent subterranean aquifer which carried fresh water to the City of London in the eighteenth century (6.22:77) (6.23:78).

The third important London outbreak, in 1952, was at the Middlesex Hospital. The nurses of the hospital have residential accommodation at John Islip House, a mere twenty metres away from one of London's most important storm relief sewers (6.24:4). Sir Donald Acheson, now Britain's Chief Medical Officer, investigated the first important outbreak at the Middlesex as a young registrar. (Most commentators for some reason ignore the earlier report by Houghton and Jones in the *Lancet* in 1942, which appears to be the earliest account of ME in the United Kingdom (6.25:135).) He was puzzled to find that not only was the outbreak confined to the nurses and scarcely affected either the patients or the doctors there, but that the first two cases which presented could never have been in contact with each other, ruling out direct case-to-case transmission.

All victims were young females between 19 and 32, he reported, with an average age of 22, resident at the day- or night-nurses' home during the previous three weeks, though one sufferer had only arrived three days before. What puzzled Acheson was that the first two cases presented on the same day; a junior night-nurse working in the general hospital, and a senior night-nurse in the private block. The epidemic started on 7 July 1952, just after a summer flash flood had surcharged the nearby sewer, delivering its debilitating fields into the residential dormitories.

There had been no contact between the first two victims, so Acheson naturally, but in my view erroneously, surmised that either a missing case had preceded the first two, or that there were two distinct external sources of infection, or that a symptomless carrier or carriers were already present within the community. Food and water he had already ruled out as a source of infection, and the subsequent report clearly documents his bewilderment (6.26:135). Perhaps this belated explanation will reopen the enquiry!

The knowledge that there are influences underground which

can affect human beings is not new; the Chinese Emperor Kuang Yu is said to have published an edict which ordered the probing of building sites in respect of underground water currents in order to get rid of noxious influences. No ancient site is ever found built on such subterranean aquifers. The medieval builders would test their sites first by putting sheep there and observing if they avoided the area.

Piezoelectricity and Geopathic Zones

Piezoelectricity, as well as water-generated ion flows, can cause chronic ill health. Piezoelectricity is generated when a material is squeezed; where for example, two tectonic plates of the earth's crust meet. A simple demonstration is the piezoelectric spark caused when a crystal is tapped, a phenomenon commonly taken advantage of in relighting gas appliances. It is not sensible to build on a site where a geological break is evident. You can detect this when the volume or vision of your television set is not up to scratch, since the radiation is in the VHF range. The result of building in such places, claims Lotz, is disturbance of blood circulation, insomnia, rheumatism, tensile nerves, or even cancer.

Given that the population of Britain has doubled this century, and there has also been a trend towards smaller families and home-ownership, the number of houses now built on geopathic zones has probably increased, with concomitant increases in illness. These houses will never give good health to their occupants. Lotz also recognizes the ill-effects of artificial electromagnetic fields, and cites at least one death in a house irradiated by fields from a transformer. He has some suggestions to combat such radiation, based on research by Dr Robert Endros, who found that its frequency was in the radio and microwave range (6.27:88). Using the heterodyne principle, his equipment will sense the incoming waves and generate an exact reversed copy of them, so as to neutralize their effects. Similar but simpler and less expensive equipment is manufactured by The Dulwich Health Society, though these devices seem to be limited to power frequencies (6.28:105).

Professor Herbert L. Konig of the Technical University of Munich recounts how this principle too has been realized or acted upon, subconsciously at least, for centuries. Superstition holds that the common horseshoe brings good luck if hung by the door of a house; Konig points out that the horseshoe's shape makes it into an unintentional open oscillating circuit, with a wide natural resonant frequency around one GHz. This can then act as a stabilizing carrier wave, protecting the resonance of the brain's rhythms. Such a frequency is close to the ratio and microwave ranges discovered by Lotz and Endross to emanate from geopathic zones. One GHz frequency implies a wavelength of about 21 cm, which is the range of hydrogen resonance, says Konig. This will reinforce hydrogen bonds and so help prevent DNA fracture.

Such parameters as blood sedimentation rate, he says, are also shown to change when people stand or sleep in geopathic zones, leading to heart disorders and other abnormalities of the blood (6.29:147).

There are visual clues which can betray geopathic zones, such as consistently damp walls where the damp rises only in one place, or where there are frequent lightning strikes. Some animals seem to prefer geopathic areas, while others avoid them; cats like to sleep in high electric fields, whereas dogs do not, it is said. Perhaps that is why today almost half the cats in London have serious immune deficits (6.30:109).

Neutralizers apart, Lotz claims that some protection from microwaves, natural or otherwise, can be achieved by judicious use of materials. Thick concrete keeps out such waves better than porous brick, he says. One is immediately put into a quandary by this view which contrasts with that of Hartwin Busch; either concrete is good for you as a result of its protection, or bad for you by preventing ion flow! Here is Lotz's table:

Material	*Protection (%)*
Asbestos-covered wooden roof	74
Claybrick tile roof	17
Copper sheet roof	9
Aluminium sheet roof	6
Aluminium-covered thatched clay	79
Concrete roof	55
PVC or linoleum	82

Source: Do You Want To Live Healthily? K.E. Lotz, Paffrath-Druck KG (1982).

Glass, says Lotz, is particularly unfavourable. I find that strange, since glass, unlike plastic, keeps out ultraviolet radiation. However, Lotz is more concerned with 'biological' microwaves which he argues are beneficial. These confusing views may not after all be mutually exclusive; it may simply be a matter of frequency 'windows', whereby frequencies quite close together may be alternately noxious and beneficial.

We all need the beneficial influence of the earth's natural rhythms, says Lotz. Buildings which shield us from it are malign. Layers of denser and less permeable materials–successive layers of concrete floors in high rise apartment blocks for example– break up and concentrate such radiation into pernicious concentrations. Curiously the infamous 'orgone accumlator' of Wilhelm Reich worked on a similar layered principle, with alternating layers of wood and metal. So did the Ark of the Covenant. Lotz advocates a return to natural floorcoverings like wood planking, in order not to disturb the earth's natural and beneficial radiation. He notes that after World War Two many German houses had to be rebuilt quickly, and the choice of concrete panels and metal windows in preference to wood and lath and plaster set the scene for ensuing national ill health. (Perhaps in consequence West Germany became world leaders in their appreciation of these biological effects; they were the chief guinea pigs!)

While both Lotz and Busch advocate a building designed to admit natural radiation, it is difficult to see how this squares with the increasingly high background incidence of microwave, radio and power frequency energy which is everywhere around us and known to have adverse health affects. Developing a building which satisfies both requirements does not appear to be easy.

However, I find it intriguing that this conundrum has to some extent been solved in the design and construction of the Great Pyramid of Gizeh. Not only is the pyramid itself set precisely on the earth's north-south axis, as if to minimize interference with the natural geomagnetic field, but until vandalized by succeeding ages, its sides were apparently made of exceptionally polished marble, and placed so tightly together that not even a penknife could be inserted between the blocks (6.31:155). This suggests some precaution against irradiation from any direction above the building–any such waves would be deflected away by the polished surfaces. Moreover, the pyramidal shape would draw massive amounts of negative ions to its apex, as has already been explained, and in doing so these would pass through the building to give an almost entirely negative ion atmosphere in its interior chambers.

As for the King's Chamber itself, deep in the centre of this massive structure (which is still the world's heaviest building), what might be the significance of those huge blocks of granite above it, capped with two even larger blocks, again angled as if to deflect some incoming irradiation? Could this also be a protective device? There is a single channel leading upwards to the Chamber from the depths of the earth below the pyramid, known by convention as the well shaft. Since the shaft is diagonal to the perpendicular, this can never have been its purpose, though there is water at its base. It would however serve admirably as a wave-guide for the Schumann resonances, and lead them gently into the chamber, without collision with other directed beams from above or below.

If one were to construct the ideal electromagnetic 'clean room', the solution would be very similar to that ancient structure, whose significance is lost in time. We may still have lessons to learn from our forebears appropriate to the new electric age in which we live. Perhaps, one might speculate, it has all happened before.

7. Terminal Illness: How to Live with Computer Screens

It is a violation of the most fundamental human rights to impose risk of death upon individuals without their consent. Human rights should not be sacrificed to the pursuit of a healthy economy, affluence, progress, science, or any other goal.

Dr John Goffman, *Radiation and Human Health*, 1981 (7.0).

Today, the computer screen has become a familiar part of modern life. Its ubiquitous presence permeates homes, replacing the manual, electric and electronic typewriter, or doubling as computer games-machine for bored children. Pioneering companies like Sinclair, Amstrad, Commodore, and others have opened up large consumer markets almost non-existent in the mid-seventies.

Myriad computer software programs, from the archetypal *Space Invaders* to complicated spread-sheets and accounting packages, have helped persuade even the most resistant souls to familiarize themselves with the language of the computer programmer. Children rarely actually learn to program the machines themselves. Nevertheless its vocabulary is infiltrating into our ordinary speech; 'software', 'downloading', and 'hard copy', are but a few examples of computerese retrovirally inserted into the English language.

The Growth of VDUs

In 1985 there were around 13 million VDUs (VDTs in North America) in the United States and Canada, but by 1990, these had grown to over 30 million in the United States alone. In Britain the number is now between 5 and 8 million. Unlike the television, whose somnolent spectators sit perhaps 10 feet from

the screen, a VDU operator's head is less than 30 inches from
its cathode rays. The first indication that these devices were
irradiating at a distance can perhaps be linked to complaints
that home computers were affecting television reception, not
only in the user's own home, but even next door. (I have
observed one extreme case where the neighbour's sensitivity
enabled her to say exactly when the computer games machine
next door was switched on.)

In the office, the cost efficiency of the VDU is undisputed.
Ursula Huws records that at the Grattan mail order warehouse
in Bradford, England, full-time staff were reduced from 1,000
to 550 and part-timers from 100 to 50 over a period of a few
months, when a new computerized system for dealing with
orders was introduced in 1979; this was despite an increase in
the volume of business (7.1:136). If computing was cost effective then, it is now even more so, with prices for hardware much
lower. The machine which types out this book costs only 10
per cent of the same performance machine of a decade ago,
and the publishers receive a diskette which does not need to
be manually re-keyed and composed into book typestyles.
Moreoever its mean time between failures has diminished considerably, and no longer does the occasional infuriating voltage variation lose programs or data the way it used to.

Software packages have improved, thanks to the efforts of millions of working hours over the years, and are available off-the-
shelf for hundreds instead of thousands of pounds. The development of OCR (optical character recognition) means that documents no longer need to be laboriously input by a keyboard
operator. Before long, terminals will be actuated by voice alone,
and respond in the same vein. Finally, advances in technology will substitute liquid crystal displays for the cathode ray
tube itself, eliminating any outflowing electric fields. Smaller
versions of the VDU have helped expand its domain into the
retail point of sale; supermarkets, pubs, bars, and online
banking-system cash-dispensers all incorporate their own video
display.

The benefits of this technological advance are evident. Supermarket print-outs are fully itemized, cash is available at any
time of the day or night, machine operatives can control the
production process from afar, the brewery chains can monitor
sales down to the last gin and tonic or whisky sour, and the
supermarket check-out clerk's effort is reduced from a weari-
some mental effort to a few deft hand flicks over the laser reader.

It is a short step from there to the fully automated supermarket. In fifteen short years there has taken place a fundamental and probably irreversible change in the way advanced nations buy and sell their daily bread, not to mention the sophisticated dealing consoles which are their counterparts in the financial world. Ursula Huws believes these changes were initiated by a slow-down in post-war growth which forced management to turn attention to the cost efficiency of their businesses. Doubtless the vagaries of staff attendance and the possibility of strike action had also their part to play.

VDU Emissions

The VDU operates by firing electrons at a phosphorus-coated screen and deflecting them en route by means of magnets to form the required image. These emissions are at many frequencies, including ELF, visible light, UV, and even soft x-rays. The screens are designed to filter out soft x-rays, a task they do not always accomplish over their entire life. Soft x-rays can also emerge from the back and side of a machine, and since such VDUs are often used in reception areas, waiting clients may find themselves gently irradiated while waiting for their appointment.

Tests have found maximum emissions of 0.3 milliRems (mR) per hour from VDUs. Although not especially alarming in itself, long-term exposure over a working year at this level constitutes a dose almost exactly equivalent to the maximum recommended exposure for therapeutic x-rays. (7.2:73).

A 'Rem' stands for 'Roentgen Equivalent Man'. When an x-ray's energy is emitted it is measured in Roentgen, but since different substances absorb greater or lesser amounts, it was necessary to express the amount of absorption rather than simply the energy emitted, so the term 'rad' ('radiation absorbed dose') was coined. At x-ray strengths one rad is equivalent to one Rem, but at other frequencies they will not be the same. One Rem is equivalent to ten milliSieverts, which is the international unit of dose equivalent. One Sievert is calculated by multiplying the absorbed dose in Grays by a quality factor for the particular effectiveness of the radiation. Before you ask, the Gray is equivalalen to one joule of energy absorbed per kilogram of matter such

Magnetic Field

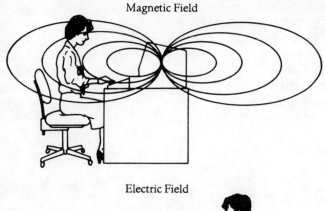

Electric Field

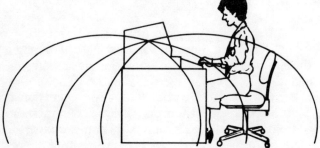

Figure 7.1: The magnetic and electric fields from a VDU.

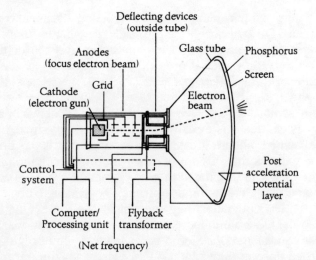

Figure 7.2: How a cathode ray tube works. Would you sit in the
firing line?

as body tissue. One hundred rad is equivalent to one Gray (abbreviated to Gy), but the Gray is now replacing the rad in common parlance, and the milliSievert is replacing the Rem.

Understand? No? I'm not surprised! The history of radiation measurement is full of changed nomenclature and confusing terms. It is also a history of falling exposure limits, which need little explanation:

Year	Maximum exposure limit
1900	10 Rems per day (effectively no limit)
1925	52 Rems per year
1934	36 Rems per year
1950	15 Rems per year
1957	5 Rems per year
1989	under review again!

The awful truth is that we still don't know what limits to impose. Long-term studies are showing that the effects may not become apparent for decades, and may even not show up until the next generation. Since we have only mapped a percentage of the human genome, damage from ionizing radiation and non-ionizing radiation alike may have caused mutations yet unrealized in our species.

Furthermore, it may not simply be a question of Rems or milli-Sieverts; repeated doses of low intensity over a period of time are currently thought to be more damaging than an equivalent dose in a few short bursts of radiation. The guidelines of Hiroshima and Nagasaki are unreliable, in that the explosions wiped out all the high-voltage electricity systems, which might otherwise have had a synergistic effect. Today we have much higher background levels of radiation than in the early parts of this century, even though only a fraction is caused by nuclear installations themselves.

Hundreds of thousands of people each year visit Sellafield's multi million pound exhibition, and are given glossy booklets which include the table below:

Radiation Source	Per cent
Internal (eating, drinking)	17
Terrestrial (gamma rays)	19

Cosmic rays	14
Radon and Thoron	37
Medical (mainly x-rays)	11.5
Other (including nuclear)	1.5

(from *Nuclear Energy: Don't Be Left In The Dark*, BNF plc, 1988) (7.2a:34).

These sources nevertheless in total amount to some 135 to 185 mR per year, which is only one-third of the annual dose which VDU operators might receive from the machines which provide their livelihood.

Microwave Irradiation

As well as x-rays, VDUs emit radiation in the UV and non-ionizing microwave frequencies. Skin cancers have been known for decades to result from UV radiation, but the effects of microwave irradiation of the head are only just being realized. (The long term follow-up by Baruch Modan has already been mentioned.)

A study by Swicord and his team found that the DNA helix will resonate when irradiated by microwaves (7.3:240). This means that it is in danger of fracture and possible mutation. I have already pointed out that the mutational forms of leukaemia, the myeloid group, are found most often near radio and television transmitters near Sellafield. Wheeler also found a correlation (7.4:265).

Other early researchers like Webb and Booth (7.5:260;261), Don Justesen (7.6:140), and Dumansky (7.7:82), began finding even more sinister effects. One doctor, William Ham, was moved to advise that 'both the retina and the lens should be protected throughout life from both blue light and near-UV radiation' (7.8:73). His admonition followed the discovery that the retina is more sensitive to injury when exposed to light in the blue portion of the spectrum–the blue light that we see in fluorescent lighting and VDU screens.

Mays Swicord of the United States FDA found that DNA absorbed 400 times more energy than the surrounding salts (7.9:240), thus confirming work done by S.J. Webb in the fifties and sixties (7.10:261). Don Justesen, who now works for Kansas City Medical Center, uncovered an even worse horror. We have a barrier, known as the blood brain barrier, or BBB for short, which keeps out contaminating agents from our cerebrospinal fluid. Microwaves, found Justesen, damage this barrier, making it permeable.

In Poland Yuri Dumansky and his research team were investigating the effect of microwave and related radiation in built-up areas in the early seventies. He found that blood-sugar levels rose and there were changes in carbohydrate metabolism as a result of exposure to densities as low as 100 to 1,000 microwatts. In the United States Deitrich Beischer also found blood changes, notably in what are known as triglycerides (a form of blood sugar), as a result of exposure to extra low frequency radiation for just one day, but his work was questioned, and his research funds terminated (7.11:22). Meanwhile Dumansky had extended his work and found that changes in liver function were another corollary of microwave exposure (7.12:50).

Another early researcher into the effects of microwaves on our eyes, was Dr Milton Zaret. He noticed that unusually large numbers of radar technicians were coming to him with post-lens cataracts, and eventually pinned it down to the microwave radiation to which they were exposed (7.13:272). The same sort of cataracts showed up in pilots, air-traffic controllers, radio operators, and VDU operators. 'Except for the VDT operators', says Bob DeMatteo 'most of these other groups have been awarded compensation. In most cases the exposure levels were below the official Canadian limits of 1 mW per square centimetre (7.14:73).

In 1987 Arthur Guy was awarded the BEMS accolade of the D'Arsonval award, after a lifetime of bioelectromagnetic research. During his role as adviser to IBM he discovered that even exposure as low as less than half a milliWatt/cm^2 could cause malignant tumours in rats (7.15:111). He was not alone in this finding; Stanislaw Szmigielsky from Warsaw, Susan Praunitz and Charles Susskind of Berkeley, California (7.16:209), and Bill Morton from Oregon all found correlations between extremely low (nanoWatt) levels of broadcast radiation and leukaemia, as did John Lester and Dennis Moore from Kansas (7.17:156).

Changes in brain-wave patterns are also induced by the kind of radiation coming from a VDU, even as low as power densities of 20 microWatts/cm^2. The result is to induce behavioural symptoms such as sluggishness, depression, and inability to concentrate. These may subsequently turn into myalgic encephalomyelitis and chronic disability.

How does the VDU produce this deadly cocktail? Microwaves are emitted by the two sweep oscillator circuits at the back of the cathode ray tube, and are therefore at their strongest at the flyback transformer at the rear, rather than at the front of the screen; thus irradiating nearby workers. 'A simple test for radio frequency radiation', states Ursula Huws,'is to hold a transistor radio tuned to a VHF station near your VDU. As you move it around, you can tell where the emissions are strongest by listening to the interference to the sound' (7.18:136).

Sweep oscillators are of two types, a vertical deflection coil, and a horizontal. The vertical coil moves the beam of electrons from top to bottom and up again sixty times a second, and thus produces a pulsed extra low frequency field of the kind which Ross Adey found could damage cerebral tissue (7.19:6). The horizontal deflection coil moves the electron beam left to right and back again 15 to 20 thousand times a second and thus pulses out a VLF field at 15 to 20 kHz, which is roughly the resonant frequency of the cranium of an infant *in utero*.

There is no doubt that these frequencies have biological effects. One of these is RF hearing, where low intensity radiation affects the cochlear myosin filaments connecting the inner hair-cells of the ear with the neural pathways to the brain (7.20:92), and induces a sensation of noise. This may well persist in the head of the person exposed long after the 'sound' has been switched off (7.21:93). De Matteo cites several studies of the effects of the two kinds of sweep oscillator:

(1) Ontario Hydro Research Divsion, 1982. The peak field strength at 30 cm from the screen was up to 170 V/m from older types of VDUs and the average was 55 V/m. I have found this strength (at power frequencies) causes hyperactivity in young children, and so it may cause the same effect in computer-games crazy youngsters at VLF frequencies, which are more energetic (7.22:119).

(2) Dr Hari Sharma of Ontario's Waterloo University found averages of 1,200 V/m at the screen's surface and 25 V/m at a 30 cm distance. He conducted a study among the VDU operators in the accounts department of the Surrey Memorial Hospital at British Columbia, where from a total of seven pregnancies there had been three miscarriages and three babies born with birth defects. Dr Sharma found that the VDUs in the accounting office had significantly higher pulsed very low frequency emissions than those in the medical records office, where preg-

nancy outcomes were normal (7.23:223).

The results of Sharma's study actually led to a change in the Ontario laws, excusing pregnant operators from VDU duties. No such regulations protect British women at this time. Britain's Health and Safety Executive published this statement in March 1987, well after Sharma's and other studies had been reviewed:

In the population as a whole one in five to one in ten known pregnancies ends in miscarriage. Occasionally, however, a group will show a much higher or lower level than these average figures. Higher levels have been reported among groups of VDU operators, but investigations show that they are not peculiar to VDU work and are to be expected on the basis of statistical chance rather than the result of VDU work. The very latest research studies have not been able to show a link between miscarriage or birth defect and VDUs (7.24:123).

(3) Dr Arthur Guy, who advises IBM, found a 10 V/m electric field at 25-30 cm from the screen, though older models were as high as 50 V/m. Guy also showed that fields at these intensities still induced electrical currents in human tissue which were biologically significant, and expressed concern about their long-term effects on operators. Dr Guy also showed that the waveforms from the VDUs were not of the gentle sinusoidal type, but sawtooth, a shape known to be more biologically damaging (7.25:111).

Notwithstanding these research results the Health and Safety Executive say in the same pamphlet as above:

As a VDU ages it is more likely to develop faults such as drift and jitter of the images on the screen. It is possible that the brilliance control will need to be turned up but this does not mean that there will be any increase in the other non-visible radiations. Machines should be serviced if there is deterioration in the visual image but there is no need for periodic radiation checks (7.26:123).

About the same time as the Health and Safety Executive were preparing their pamphlet (recently reprinted unchanged), Dr Marilyn Goldhaber of the Kaiser Permanente Medical Care programme was conducting a case-control study of pregnant

women from Northern California. This group is called in medical terms a cohort, and its progress is monitored to see what
happens to it over time. Dr Goldhaber was a little concerned
that other studies in Finland, Sweden and Canada, as well as
the United States, were missing early miscarriages because of
the way they were set up, and that the conclusions which they
had come to, namely that it was probably safe to use VDUs
while pregnant, were therefore wrong. Originally, the cohort
had been established in 1983 to test for the effects of the pesti-
cide malathion.

The research team sent a questionnaire to over 1,800 of the
original cohort, and eventually got answers from 1,583, over
80 per cent of them. The questionnaire included routine questions about their work and whether they used VDUs and if so,
for how long each week. Other questions were about their use
of tap water, bottled water, caffeine, alcohol, and cigarettes. When
the results were analysed, it was found that the estimated relative risk with any form of VDU work was only 1.2 to 1.0 and
not enough to get excited about. But when they looked at the
risk for women working more than 20 hours a week on the
machines, they found the risk rose to 1.8. In other words the
chance of miscarriage was nearly twice the normal expectation
for that group. Even then the risk of birth defect was not quite
so high, at 1.4.

It could be that the women who had suffered miscarriage over-
reported (i.e. blamed) their time spent at the console. This is
one of the dangers of using questionnaire techniques in
epidemiology. But there were so many other questions on the
sheet that this is unlikely. Another possibility is that the women
exposed to VDUs for longer were less mobile than the others,
and the lack of exercise was a contributing factor.

Since the Goldhaber results were published no substantial
further study has yet emerged, despite their conclusions:

Our case-control study provides the first epidemiological
evidence based on substantial numbers of pregnant VDU
operators to suggest that high usage of VDTs may increase the
risk of miscarriage. The implication of this finding is yet
unknown. No biological mechanism has been postulated, nor
has a clear pattern of risk been observed across all occupational
categories. Our data do however suggest that need for further
investigations (7.27:102).

What Can You Do?

Obvious, costless and immediate steps to avoid these hazards begin with siting the VDU itself so that it does not irradiate nearby personnel. Are you sitting with your back towards someone else's machine? Electric field strengths attenuate with the square of the distance, and magnetic fields with the cube. Thus, if an electric field is 100 V/m one metre away from a source, it will only be 25 V/m at two metres. Magnetic fields are more complicated than this because they have several components and vary in strength during any period of time. Given the educated guess that an E-field of less than 10 V/m is relatively safe, and taking a lead from Dr Guy's work, at two metres the average electric field from a VDU should be only about 8 V/m. Of course averages are always composed of higher and lower figures, and the only real way to know is by means of a field-measuring device. (The manufacturer of the VDU may be prepared to take these measurements *in situ* and advise you–especially if there is a prospect of more equipment sales). The other uncertainty is the age of the VDU itself. Older VDUs are likely to radiate more.

Bearing in mind that your own body is the other component of the potential difference between the VDU and you, it is a simple precaution to make sure you yourself are properly earthed; wearing natural leather soles, slipping your shoes off, or even a surreptitious wire clipped to the ankle and earthed are all good ways of ensuring that your electric charges do not build up when working. Do not forget the hook-up when about to get up and leave! The VDU should also be properly earthed for the same reason.

Electric fields are relatively easy to shield. There are a number of anti-radiation devices or screens on the market, from which I recommend the No Rad Corporation's dB 60. It claims superiority of protection over others in the market, and consists of a black-coated very fine copper-mesh to catch the electrons which are then led to earth by a wire. Some screens have plastic cases, so it is never a bad idea to earth the screen separately.

Whichever protective screen you buy, make sure it includes radiation protection, and is not simply for glare. (The latter are obviously much cheaper, but this may be a false economy if you are then incapacitated for a year or so by radiation, as some VDU operators are.) Writers working in isolation are particularly vulnerable to financial calamity from VDU irradiation.

The back and sides of the VDU can also be shielded. There are a number of suppliers in the United States, Canada, and Britain, who can supply metallic fabrics, some adhesive backed, to cover the machine.

Another common health hazard for VDU workers is repetitive strain injury (RSI), or tenosynovitis. Bearing in mind the possibility that radiation may be causing more discomfort in the myosin filaments in the muscles of the hands than the physical activity itself, it is worth putting a metal screen between the cathode ray tube and the keyboard, low enough to see the CRT over the top, but high enough to keep off any electric fields. Earthing this screen will improve its performance as a Faradic barrier.

I discovered an old television set with a bakelite case, and in disembowelling it in order to put in modern guts, I was surprised to see that its interior had been shielded with aluminium foil. I would not however recommend that you tamper with the inside of the modern VDUs! Better to make sure that all new VDU purchases are of radiation-reduced apparatus, and though this aspect is not often specified on the sales literature, the manufacturer will tell you if you ask. Ericsson's VDUs were designed in this way, and their present owners, the Nokia Group of Finland, who are one of Europe's leading VDU suppliers, have continued and extended this policy. By using a higher than normal refresh rate (70 Hz in character mode) the ELF frequency is kept away from frequencies near those emanating from the brain's own carrier frequency of around 4 to 20 Hz. Taxan is another manufacturer producing low-radiation screens. Nokia's and Taxan's declared radiation levels are:

	Nokia	*Taxan*
X-ray	less than 0.5 mR/hr.	n/a
Magnetic	less than 25 mT/s	less than 23 mT/s
Electrostatic	less than 500 V/m	less than 500 V/m

Conventional monitors give out static electricity of 6 kV, compared to the 0.5 kV above; magnetic induction of 88 mT/s (against the less than 25 nT/s shown above), and flux density of 270 nT, compared to about 50 nT with the Nokia and Taxan screens. Nokia are now also producing liquid crystal displays with almost non-existent electric fields, which is the way all the technology is going (especially since the footprint is so small, and takes up minimum desktop space). IBM has lately started patenting low radiation screens and offering low emissive products.

Another means of E-field shielding is to use nickel-based acrylic paint. The risk is that the paint cracks after drying, and it is more difficult to ground. Possibly also it is not cosmetic enough for smart modern offices! Magnetic-field shielding by contrast is notoriously difficult if not impossible to achieve, though some VDUs have been balanced enough not to emit magnetic fields (or H-fields as they are sometimes known). We might learn from nature here; as everyone knows our blood contains iron, which is of course magnetizable. I have observed that the complicated structure of haemoglobin seems designed to prevent magnetization of the haem's ferric material by curling the globin around it like four horseshoes, and then lining up the ends so that the opposite polarities are facing each other. I do not think that anyone has ever considered the complex structure of haemoglobin as designed for that purpose, but if so it would be in accordance with CMR theory; electrified or magnetized blood would fog the brain's signals.

Maybe something like that happens with the new kind of immune deficiencies we are now experiencing. Certainly for some reason the haem's iron, amounting to 22 mg in weight in the average body, is retained, and very little relinquished even though iron passes through the body in much greater quantities than that every day.

A magnetic field is measured by a coil of a known number of turns of wire connected to a device which measures the current in milliTesla or nanoTesla. (Again, there are variant measures such as Oersted, Gauss, and gamma but they need not concern us here.) The magnetic field and the electric field when multiplied together usually give the total power density of an EM source. Which of these two components of EM energy is more biologically damaging is still the subject of debate and research, complicated because the magnetic field can pass through most substances, and re-induce an electric field as it changes in intensity.

There are other, less orthodox techniques claimed to reduce cathode-ray tube radiation, which I will mention because there appears to be an element of truth in them. Even science should be entertaining. The first device is a large quartz crystal placed on the top of the VDU! I have observed that this does seem to absorb the E-field even though it is not directly in the way, and presumably it absorbs electrons or ions which should presumably be washed out in running water every day. (The running water promotes ion exchange and re-balances the system.) How such crystals can pull the ions from the front of the screen beats me, but the instruments show that they do. Early radio sets contained crystals which were capable of accepting the frequencies imposed on them, which were subsequently amplified.

Another bizarre method is to place a certain cactus, Cereus Peruvianus near the screen! Claims for its effectiveness come not only from the United Kingdom distributor, Cirrus Associates, but also from the Institut de Recherches en Geobiologie at Chardonne, Switzerland, whose spokesperson Blanche Merz, in October 1986 said:

We have distanced ourselves from a large number of protective devices, especially those which have had a lot of publicity. And yet the concern about coming to grips with the seemingly aggressive emissions from television and computer screens besides the background radiation has always remained with us.

Even if coincidence is said not to exist, the ancient wisdom of nature has enabled us to make an important discovery three years ago on the high plateaux of Mexico: a plant is an effective antidote for the intelligent inventions of the human brain in those cases where technology does not concern itself with damaging side effects.

We have carried out a lot of tests in the meantime, so as not to come to any rash conclusions. For the greatest reliability the practical tests can be carried out using the Bovis Biometer. For two years certain people in the United States have conducted blind tests on Wall Street; an approximately 40 cm-high cactus was placed next to every computer screen. The employees who used to suffer from headaches and tiredness felt, both physically and mentally, in top form afterwards (7.28:176).

The Bovis Biometer mentioned is even more bizarre than the cactus, since it was invented by the man who discovered pyramid energy and managed to patent his device for sharpening razor blades under them. I would personally be interested more in the blind tests than the somewhat subjective measuring methods of Bovis! However one should keep an open mind on these things. After all, the humble aspirin is nothing more than a concentrate from the willow tree, and many pharmaceuticals are of plant origin, so why not EM protection devices?

One can at least admit that cacti, which spend most of their life under the gruelling sun — our most important source of electromagnetic energy — might have developed techniques for absorbing its excesses. I look at the morphology of cacti in a new light these days; what *are* those little aerial-like spines for, anyway?! Are they emitting negative ions in compensation for the positive-ion emissions from the VDU or the sun?

No one seems to be trying to make capital out of this little plant, and even if it doesn't work, it can do no harm to have a cactus by your screen. I have one here as I write, and it is quite comforting! However, a quick check with the meters tells me that the E-field a foot away is about 10 V/m and is modified down to 8 V/m by the cactus (as it would be by any opaque object), while the H-field's vertical component is fluctuating around 50 nT minimum whether the plant is there or not.

Another curious device or devices, for this is only one representative of a whole class of neutralizers, is called the Cosmoton. Originally made in Germany, by Ordo-Stiftung, a non-profit making concern, it is marketed in Britain by Hildegard Pickles of Leeds. The Cosmoton, like the Harmotron and others, is worn as a pendant. Since it takes between three months and a year to show its full effect, measurement of effectiveness is not easy, but it was apparently developed by the same Dr H. Palm mentioned elsewhere, and originates with the multiwave oscillator developed by Lakhovsky against cancer. It seems to consist of seven horseshoe-like rings of different gold and silver metals like the planets. It requires no battery, being 'charged from the cosmos'.

I do not intend to sound mocking; the ancient Greeks wore crystals round their necks in similar fashion; our word 'cosmetic' comes from that practice, which was said to keep one in tune with the universe or cosmos. 'When the cosmoton is worn on the chest or placed on the television or VDU screen', says the sales pamphlet, 'it helps by protecting children and adults in tolerating the radiation of the screen far better without ill-effects. ...The diameter of the effect is about two metres' (7.29:204). Again, I doubt whether such a device can do any harm, and might be a fairly attractive adornment. It looks like a miniature multiwave oscillator, invented by Lakhovsky after an idea by the great Tesla.

It may appear to be stating the obvious, but another form of protection is simply to avoid working at the VDU for more than two hours a day. No research has ever found that such levels of exposure are hazardous. A quiet and informed discussion with a reasonable and caring boss could well lead to altered work schedules designed to accommodate such exposures. That way no one gets saddled with over-exposure at the possibly dangerous levels of twenty hours and more.

Another device on the market is a protective jacket called the Microshield. This jacket attracted some flak from the National Radiological Protection Board on British television, but the attack itself was unfair in denouncing the manufacturers' claims; they had not claimed that the jacket could stop magnetic fields, which is what the NRPB demonstrated. On the other hand the wording in the firm's literature was imprecise, because they referred to electromagnetic fields, when they should have said electric fields. Accordingly they deserved a certain amount of critical treatment. The jackets do not protect the face and more importantly the brain of the operator, which is where I personally think the trouble starts; but they will certainly reduce the radiation levels from E-fields. However, the crucial question is, is it the E-field or the H-fields which do the damage? Or both? The NRPB ought really to be finding out the answers, rather than just quoting a few selected negative reports.

From Switzerland comes A-Nox, another curious device which consists of two translucent spheres containing 'rare earths', which are placed diagonally on either side of the screen frame. They need no power supply. The constituent elements are claimed to enter into resonance with the electromagnetic field of the cathode tube of the screen and 'generate an energy field acting as a magnetic shield for the operator's protection' (7.30:239). Jacques Surbeck, who markets the device world-wide, produces test results to support his product's efficacy. Unfortunately he omits to say who carried the tests out, a vital piece of information I would have thought. If the testing laboratory were found to be acceptable, the results would certainly justify the use of A-Nox devices, despite their price. However I could not find any diminution of the E- or H-fields with my own equipment.

Perhaps the most bizarre protection device of all those on offer to the concerned VDU operator is a plastic card, called the

Charge Card. This looks like a bank credit card with some small circuits on it. The user is supposed to charge the card up with his or her own vibrations, though I cannot see how it works from inside a handbag or wallet!

There are also various kinds of pulsing devices available, such as Biomag and Mecos, which are unfortunately quite expensive (and a quartz watch may do the same for a few pounds!). These claim to offer protection from noxious rays by pulsing out a steady protective EM pulse, presumably thereby acting as a carrier wave for the body's own signals. I have no opinion on them since no one has ever established blind trials of their effectiveness.

To protect the brain of a VDU operator from E-fields some kind of veiled headgear is necessary. However, fashion and human self-esteem being what it is, it is unlikely that any operator will wear such a covering until it becomes acceptable behaviour, like hard hats on a building site, welding visors, or motorbike crash helmets. The hazard is so invisible and long-term that few will be prepared to pioneer the idea. More sensible in the short-term is a pair of plain glass spectacles, which at least will shield the eyes from the UV component of the radiation, and lessen the risk of cataracts. Ferrous metal frames are not a good idea, since they concentrate magnetic fields. Noble metals are better, but so many spectacles have plastic-coated ends that a static charge could build up in the unearthed metal and might cause a headache.

Finally, one should be aware of negative ionizers as a way of limiting the positive ions discharged from VDU screens; some operators complain of blotchy faces after working on the VDU, and this may be due to the impact of positive ions on their mildly negatively charged facial skin. A negative ionizer will lessen this risk.

The biggest protection of all is simply being aware of the hazard. Believe me, there are plenty of cautionary tales to tell; for example, the president of an ME action group was an author who was smitten with the debilitating 'Yuppie Flu' after having sat down at a VDU every day for several years to write her novels. I gather that after changing the computer to a low radiation version she is now much better. Fortunately myalgic encephalomyelitis seems to be reversible, unlike the more serious immune deficiencies, which may also be the result of radiation.

The Radiation Safety Corporation of Palo Alto, California, a private firm, can supply prepaid x-ray monitoring tabs, called On-Guard which you wear for the allotted time and return for processing and analysis. Each test costs at present about $35, though there are discounts for bulk. As a way of influencing your employer to do something about possible VDU hazards it might be a good idea to try one of these tabs, especially since, according to the firm: 'A recent Food and Drug Administration Study found that one out of twelve normal-appearing computer monitors emitted x-rays in excess of the federal safety limits.' Unfortunately the limit applies only to monitors which can display a picture. Thus VDUs without graphics are outside the jurisdiction of the legislation.

Eventually the problem of VDUs may simply go away; quite rapidly a new kind of screen using liquid crystals, which gives off very little radiation, is arriving on the market. Currently more or less confined to lap-top computers, as the technology and volume gets underway, the liquid crystal screen will supplant that bulky old CRT on your office desk, and give off very little radiation in use.

8. An Unhealthful Post: Working in an Electrical Environment

Above all the Physicians of our society should be desired to give us a good account of the epidemical Diseases of the year. Histories of any new disease that should happen, changes of the old, Difference of Operations in Medicine according to the weather and seasons, both inwardly and in wounds, and to this should be added a due consideration of the weekly and annual bills of Mortality in London.

Directive to Doctors, 1657, (as quoted in *The Mathematical Science of Sir Christopher Wren*, J.A. Bennett).

In 1989, soon after the twenty-fourth mysterious death of a senior scientist working on microwave technology, the *Sunday Times*, in a rare revival of its old spirit of investigative journalism, devoted a page to asking some tough questions.

Many of these scientists had died in violent, and unexplained circumstances — death plunges from bridges, hosepipes leading from car exhausts, apparently deliberate self-electrocution — all were par for the grisly course. Others had developed tumours of the brain or suffered fatal heart attacks. What was the solution to this mystery, asked the *Sunday Times*? Were they silenced for knowing too much about the 'Star Wars' project on which some of them were working? Were the suicides yet another example of what Stephen Perry had found with people exposed to power frequency fields near high voltage power lines? Were the tumours a natural result of microwave radiation, just as had been suffered by the unfortunate staff of the United States Embassy in Moscow — the so-called 'Moscow signal'? Two of the United States ambassadors serving there during the period 1953-57 died of cancer and the third, Walter Stoessel, died in December 1986 of leukaemia, which was first diagnosed in 1975 when he was taken ill with nausea and bleeding from the eyes (8.1:246).

Research in Electrical Environments and Health

Despite the fact that more than half our present-day scientists are involved in military projects, working in an electrical environment is not a sole prerogative of the military; it has been clearly linked to ill-health by several registry surveys and epidemiological studies. Among the earliest of these was a Soviet analysis of workers in electrical substations or switchyards, where high voltage electricity is transformed down towards domestic levels (8.2:146;213). A five-person research team from the endocrinology department of the Bucharest Institute of Hygiene and Public Health, carried out a study of thirty-one young technical workers whose occupational exposure to microwaves averaged eight years (8.3:152).

Their investigation revealed that the men showed a marked loss of sex drive and sexual dynamic disturbance, within a prevailing (70 per cent) incidence of asthenic syndrome (Yuppie Flu). Nearly three-quarters of the men had less than usual amounts of sperm and the sperm itself was weak. These deficiencies could not be linked to any hypothalamic or pituitary gland imbalance, and the effect was presumed to be related to the EM exposure, since an improvement in spermatogenesis was noticed in two-thirds of the technicians after a three-month interruption of their exposure.

During the Second World War, British servicemen had also been aware of this induced spermopenia, and some would pay a shilling to stand for a while in front of the radar transmitters before departure on shore leave, to avoid putting their girl friends 'in the family way'. The rumour led to one of the first official investigations of bioeffects of EM energy (8.4:35).

About the same time as the Rumanian study (October 1974), the American journal *Aerospace Medicine* published the results of another Eastern bloc study, this time from Poland, of 841 males between 20 and 45 years old who were occupationally exposed to microwaves (though exactly how they did not say). Although the researchers could find no difference statistically between those exposed to high or low levels, they concluded that:

'The incidence of functional disturbance (60 per cent) is

unusually high'. Of the 60 per cent, 20 per cent revealed digestive tract complaints, and in nearly a third of these cases: 'The severity of disturbance was thought sufficient to declare the persons concerned unfit for further work in conditions of microwave exposure' (8.5:226). The staggering thing was that the power density of this exposure was miniscule, below 0.2 mW/cm^2 (low) and between 0.2-6 mW/cm^2 (high). This compares with the official American permitted exposure limit of 10 mW/cm^2. It is not really surprising that this major research effort did not tempt the United States authorities to re-examine their exposure limits; the country was already committed to higher levels as a result of post-war exploitation of microwave technologies.

Scientists like John Clark, working directly under the main radar transmitter at the Royal Signals and Radar Establishment's south site at Malvern, are almost certainly exposed to at least 4 mW/cm^2 on most working days. Nearly half of the Polish workers (180 out of 407 of those in the 'low' group) suffered from 'neurotic syndrome', whose symptoms included:

- fatigue disproportionate to effort
- frequent headaches
- sleep disturbances
- emotional instability
- inability to concentrate
- difficulties in memorizing
- decrease in sexual potency
- ECG abnormalities like bradycardia (slow heart-beat)

These are identical to the symptoms of ME, where I know of at least two cases of associated suicide. It is unlikely that the coroners presiding over the inquests of the mysterious deaths of our microwave scientists were acquainted with this East European study.

In 1982 Dr Kjell Hansson Mild of Sweden's National Board of Occupational Safety and Health, and Ake Oberg from Linköping University published a critical review of the neurophysological effects of EM fields (8.6:179). They noted that the low-level radiation characterized by the 'neurasthenic syndrome'–Yuppie Flu under yet another name–had already been reported by Baranski and Czerski as early as 1976 (8.7:15), and went on to record that the latest index of publications concerning the biological effects of EM fields lists no less than 3,627 articles in the world's scientific literature.

Accordingly they confined their own review to papers concerning neurophysiological effects, and cited a mere 58 references by further limiting the paper only to non-optical frequencies. They divided these into occupations exposed to EM energy above 300 MHz, e.g. radar equipment, broadcasting transmitters, antennae for UHF-TV, and microwave diathermy apparatus; and those below (at which point the wave length is about one metre) when practically all radiation takes places near the radiation source within one wavelength, and this space–called the 'near field'–where the wave's form is not fully developed, is characterized by components which decrease rapidly with increasing distance:

In this frequency range the main sources of radiation are plastic welding machines, glue dryers, short wave diathermy apparatus, and broadcasting equipment for FM radio and VHF television. The field strengths near these range from over 1000 V/m and 2 A/m close to the antennae down to 10 V/m and 0.1 A/m one or two metres from the source (8.8:15).

At still lower frequencies, down to about 100 Hz, magnetic fields dominate occupational exposure to welders and steel workers and a variety of heating applications jobs. Finally there are the power frequencies, experienced near power lines and in electrical substations.

One of Ake Oberg's earlier experiments (in 1973) showed it was possible to induce muscle contractions in a frog by the rapid switching of a magnetic field across the nerve perpendicular to the long axis (8.9:191). (No wonder ME sufferers wake up feeling muscular exhaustion.) This effect acts directly on the nerve, and not on the muscle, a fact which Gengerelli and Holter had established way back in 1941 (8.10:98). Others, like McRee and Wachtel, had continued this work and concluded that the observed effect is based on interference with the long-term regulating mechanisms such as the maintenance of ionic concentration gradients across the cell's membrane (8.11:175). What they are really saying is that the root of these EM effects is the disturbance of the individual cell's polarity; it has nothing to do with any thermal effect, since the intensity is too low for that.

The growing awareness of microwave and radio frequency hazard at very low intensities was not confined to the Eastern Bloc; in the mid-seventies American researchers like Ross Adey, Carl

Blackman and Susan Bawin began to report that if cells were irradiated by low frequency EM energy or even by higher frequency radiation modulated at about 10 Hz and 16 Hz, positively-charged calcium ions started to put out of the cells (8.12:32). This efflux is, I suspect, a reflection of the depolymerized myosin and actin microfilaments, without which cells seem to turn neoplastic, which leads to cancer. Adey's work was confined at first to the brain, but later studies found the same with any nucleated cells; these cells, also called eukaryotic (the greek word *kara* means 'head') because they have a noticeable nucleus or head within them, all contain microtubules. A clear indicator of any cancer cell is the absence or near-absence of these microtubules (8.13:10).

A number of occupational studies confirm the correlation between low-frequency exposure and health hazard. In 1983 Vagero and Olin, the latter from Stockholm's Royal Institute of Technology, used a new registry of 385,000 Swedish cancer victims between 1961 and 1973 to establish that the incidence of cancer in the electronics industry was higher than for industry as a whole, and in the case of pharangeal cancer, twice as high (8.14:251).

More specifically there appears to be increased risk of brain cancer among electrical engineers. Michael McDowall, a statistician with Britain's Office of Censuses and Population Statistics, found that leukaemia, particularly myeloid leukaemia, was prevalent among telecommunications engineers (8.15:172), and similar results were obtained by Michel Coleman and his team after examining patients registered with malignant disease in south-east England (8.16:54). In Washington State Sam Milham Jr found that ten out of eleven jobs involving unusual exposure to electric and magnetic fields had an increased PMR (proportionate mortality ratio) for leukaemia, and this was confirmed after taking a look at similar data in the Los Angeles cancer register (8.17:180).

One particularly clever study by Sam Milham was based on the obituaries of ham radio operators, which appear regularly under the 'Silent Keys' column in the journal of the American Radio Relay League, QST. Between 1971 and 1983 there were 296 deaths for Washington State, and 1,642 listed for California. The death certificates of over 80 per cent of both groups were inspected, and it was found that the PMR for the myeloid group was over twice that which should have occurred normally, whereas the lymphatic cancers were at normal levels (8.18:181).

Overall the incidence of leukaemias was 1.91 times what should have been expected (some 24 observed cases against only 12.6 expected). Nearly a hundred of the 280 Washington State 'silent keys' deaths (35 per cent) were of people with occupations such as electronics technician, electrician, or radio operator, whereas only 3 per cent of male deaths in the Washington death file as a whole were so employed.

One should of course be careful here: ham radio operators are much more likely to have an occupational background in the subject which they favour as a hobby.

'These findings', concluded Milham, 'offer some further support for the hypothesis that electromagnetic fields are carcinogenic.' Because the deaths were predominantly myeloid, the same findings support the view that RF frequencies at low intensity may also cause mutations, and lead in turn to cancer when the mutated cells proliferate.

A possibly typical example of radio frequency hazard is the case of Paul 'X'. He had worked since January 1982 as a process development engineer for Plessey Semiconductors at their microelectronics centre at Hollinwood, near Manchester, United Kingdom. His work, alongside thirty colleagues, was in the 'clean room', where only particles less than one micron would be airborne. Among the equipment in that room are furnace tubes–which are like long ovens–permanently switched on, and plasma etchers, through which radio frequency waves are passed; the two in effect make the set-up like a large microwave oven.

The three plasma etchers where he worked use radio frequency at different frequencies: 40 kHz, 200 kHz, and finally 13 MHz; the last being installed most recently. The issue is complicated because he was also being exposed to noxious gases like carbon tetrachloride. Another piece of equipment in the clean room is an implanter, which shoots a beam of ions at a silicon wafer. Paul was assured by the safety officer that the radiation given off was 'lower than the upper limit'.

Until early 1983 he was fit and well, cycling to and from work each day, a distance of eleven miles. He started becoming prone to viral infection–a sign that his immune system was malfunctioning–and as time went on he started falling asleep at work in the afternoon. Eventually, around April 1987, he stopped work, at the age of 47; he has not returned since. One of his teeth kept bleeding, and the doctor suggested a blood test. He had just arrived home afterwards when they rang and told him to come straight back; the condition was diagnosed as aplastic anaemia. Numerous blood transfusions followed, and on occasion he had to remain as an in-patient for several days.

Finally a bone-marrow transplant was attempted, with his younger sister as a donor. But there were problems with rejection, not initially recognized because of kidney failure. Paul was hospitalized for several months, given steroids, antibiotics, and intravenous feeding. He also suffered chicken pox, and bowel blockage, until finally a part of the bowel was removed. As a result of his disease he has now lost four stone in weight and suffered serious muscle wastage.

The consultant chemist advising Paul's solicitor said in April 1989:

There may be some mileage in following up the effects of exposure to microwave radiation. Although I am not qualified in this area, it seems to me at the moment that this offers the greater probability of being a potential cause of Mr 'X's aplasticity.

Interestingly the clinical haematologist consulted by Paul's solicitors had also noticed 'purpuric bleeding spots particularly on his feet'. This observation could well accord with CMR theory; Kaposi's Sarcoma is also first prevalent on the lower limbs and feet of AIDS cases. The original blood count showed lymphocytes down to 2.16 (normally between 1.5 and 4.0) but none of those concerned with the case at that time had considered a microwave or radio frequency irradiative aetiology.

Parental Exposure

Several of the leukaemias in the vicinity of the Dounreay nuclear power generating plant in the very north of Scotland are also the myeloid type, especially among the children whose homes are on the hillsides facing the microwave relays and United States Naval Communications transmitting towers at Forss nearby. Once again, however, the position is complicated, since although six of the childhood leukaemia cases occured in one housing estate on the Forss-Dounreay side of Thurso, this same estate is also where many employees at Dounreay live–it is even locally known as the Atomic Estate (8.19:58). (Both employees and their children seem to have suffered, but, principally children, whether of employees, or not.) (8.20:70). Thus the question of a parental exposure link arises. Other cancer cases are found as far away as Wick, Castletown, and Dunbeath, whose sufferers also work at the plant.

Perhaps the greatest concern is what happens to the subse-quent generations of a species subtly mutated by EM energy. If the male sperm or female ovum is genetically damaged, not only may the children be genetically defective, but the deficiency may not show up for several generations. Two recent Swedish studies throw some light on this question. In 1982 Kallen and Moritz obtained detailed information on 2,043 infants born to 2,018 physiotherapists in Sweden in the mid-seventies. The results suggested that mothers exposed to shortwave equipment ran a greater risk of having a dead or deformed infant (8.21:141).

The second study, by Nordstrom and his colleagues, used questionnaire results from 89 per cent of 542 employees in Swed-ish power plants in 1979 (8.22:190). Men working as high vol-tage switchyard operators appeared to report an excess of congenital malformations in their children, though this was based on few events (26 cases). Nordstrom's team divided the employees into three groups: those who worked near the live lines; those who built the lines before they were brought into commission; and office workers at the plants. There was a statistically significant decreased frequency of births among the first group, and also a statistically higher frequency of children with congenital malformation–over 16 per cent in the latest period they studied (1971-1979).

Since 400 kV lines were only introduced in Sweden in 1952, it is interesting to note that no children with a malformation born earlier than 1953 was found by the research team. 'As a result of this study', they concluded 'prospective epidemiolog-ical studies of possible reproductive hazards among people working in high voltage substations should be performed'. So far the British CEGB, for example, has performed no such research, except to monitor its workers and establish that they are not exposed to levels of EM energy above the American permitted limits (8.23:165). Reviewing the genetic defects they uncovered, Kallen and Moritz say:

The last possibility is that repeated exposure to non-ionizing radiation in the form of shortwaves (and perhaps also microwaves) could slightly increase the risk of foetal damage. Experimental investigations provide little information on possible hazards in the human situation (8.24:141).

Margaret Spitz of the University of Texas' Anderson Hospital and Tumor Institute carried out a case-control study of neuroblastoma and paternal occupation in the early eighties. Half of neuroblastoma cases occur in children under two, suggesting some pre-natal influence may be the cause of this specialized form of childhood nerve cancer. Previous research had tried unsuccessfully to relate it to petrochemicals, ionizing radiation, and other known carcinogens.

Turning the problem on its head, Dr Spitz investigated the parental occupation of 157 children under 15 in Texas who had died of neuroblastoma between 1964 and 1978, and compared them with twice as many randomly-selected controls born in the same year, and without any differences due to geography. Parental occupation was derived from the children's birth certificates, and, since few of the mothers were employed, the father's occupation was selected for study. The resulting probability ratio for children of electrical workers in general was significantly elevated, at 2.14 times; for electronics workers in particularly (6 cases, but only one control), the ratio was 11.75. Most of these latter parents worked in atmospheres permeated by radio frequency fields, in a television repair shop for instance, or in the manufacture of microelectronic components.

Most recently a new study by a Southampton team has also linked nuclear workers to genetic deficiencies in their children, once more underlining the fact that ionizing and non-ionizing radiation are both simply parts of the same continuum (8.25:97a).

Radiation Cataracts

Apart from possible long-term genetic hazard, microwave irradiation undoubtedly has a fairly immediate effect; it turns you blind–by causing cataracts behind the lens in your eyes. This has been seen both in laboratory work and in occupational conditions (8.26:273) (8.27:274). The questions are; how powerful do the waves have to be? How long does one have to be exposed to them before damage occurs? And, finally, is the damage reversible?

According to Milton Zaret, microwave cataract shows up as posterior subcapsular opacity (PSC), a long way of saying that the back of your lens fogs up. Since radio and television broadcasting and repeater towers emit microwaves, radio linesmen who erect and maintain such towers could be at risk. So could any person living or working in the beams from the towers' antennae.

In a letter to the *Lancet* in 1984, two Australian researchers, Hollows and Douglas — one an opthalmologist, the other a statistician — reported the results of examining radio-linesmen's eyes through a slit lamp. The linesmen were exposed to frequencies ranging between 558 kHz to 527 MHz — not very high as frequencies go — for most of their working lives. They compared them with the eyes of people who had never worked as linesmen, and found that the prevalence of PSC was as follows.

Worker	PSC Incidence
Linesmen	21 per cent
Controls	8 per cent

'The results of this controlled examination', they concluded 'suggest an increase in PSC in radio-linesmen that may be work-related.' (8.28:132).

Other occupations prone to radiation cataracts include operational aviation workers like air-traffic controllers. In 1977, Milton Zaret described how the cataracts start:

The earliest symptoms of radiation cataract are often transient visual problems, such as looking through a misty fog or wet glass, symptoms which can antecede objective signs by years. For example, one of the air-traffic controllers, during a five-year period between 1967 and 1972, experienced many episodes of wrongly identifying aircraft or losing others on his radar plot. Several times mid-air collisions were prevented only by the intervention of colleagues. Throughout this period numerous opthalmological examinations and neurological examination failed to reveal the diagnosis (8.29:274).

Even surgeons carrying out operations may be putting themselves at risk, according to Jacob Paz and his team from New York Medical College and other American Universities. Electrosurgical units (ESUs) are used in operating theatres for coagulating, suturing, cutting, and other purposes. As the instrument is used the surgeon's eyes and forehead are on average only 20 centimetres from the ESU's active lead. When Paz took measurements he found that the electric field intensities could be as high as 3.5 A/m–way above the permitted exposure limits of the American National Standards Institute (8.30:198).

The authors published their findings in 1987 and concluded that such electrosurgery devices should receive immediate attention to assess health effects.'Even the New York subway and a microwave radar transmitter on Governor's Island were far below those attained by ESUs,' they claimed. Of course, Paz was only concerned with ocular effects and possible cataracts, but the surgeon's brain is not much further away than his eyes, posing the possibility that such devices may also cause brain damage.

Brain Tumours

Doctor Ruey Lin went along to the annual meeting of the Society of Risk Analysts at Knoxville, Tennessee in the autumn of 1984. It was a pre-semester break with a purpose. A few days later most of his research team would be back at their desks at Maryland's State Department of Health and Hygiene. For the moment though, their concern was brain tumours, and the study they presented was an analysis of nearly a thousand death certifcates of adult white Maryland males who had died of brain tumours between 1969 and 1982 (8.31:159).

Brain tumours are of various types; melanoma, meduloblastoma, ventricular tumours, and other rare varieties including metastatic tumours which had their origins elsewhere in the body; all these types were excluded from Lin's study. Those which interested him were:

Glioma	370 cases
Astrocytoma	149 cases
Unspecified	432 cases
Total:	951 cases

The average age of death was the mid-fifties. When these cases were compared by occupation with an equal number of controls (males of similar age who had died of causes other than malignity within the same year, selected from a computer-randomized register) it immediately became apparent that the proportion of the cases who had been working in electrical trades was twice as high as would have been expected in the glioma and astrocytoma cases (2.1 times), and noticeably higher in the non-specific tumour cases (1.6 times).

In a telling sentence the researchers pointed out:

In recent decades there has been a rapid increase in the use of radio frequency in industrial, military and civilian settings. During the same period of time (1940 to 1977) brain tumours in the United States population have risen from 3.8 to 5.8 among whites and from 2.15 to 3.85 among non-whites per 100,000 persons (8.32:159).

They speculate that the gliosis — a proliferation of the brain's glial cells — might be the result of a healing mechanism gone wrong, since pulsing EM fields induce a weak electric current in bone and stimulate the repair of un-united (non-union) fractures.

The National Radiological Protection Board

Not all scientists agree that RF and microwave frequencies can promote cancers. Britain's National Radiological Protection Board (NRPB) takes the view that while thermal effects (radiofrequency burns) may be expected at very high strengths, the paucity of scientific studies showing athermal effects contrasts with 'the many more experiments and observations which provide evidence for the absence of any biological effects at electromagnetic field strengths very much greater than those recommended as safe by IRPA/INRC' (8.33:75).

This is an unconvincing argument, since the lack of funding for epidemiological research into EM effects by British authorities has been notorious. The same argument might have been used in Galileo's time to say that since he was the only person saying the world moved while everyone else denied it, he must be wrong! The official view of the NRPB is summarized in a modified version of their 1988 report to a public enquiry over the siting of a low frequency (254 kHz) transmission mast operating at 500 kW in County Neath in the Republic of Ireland. This transmitter was some seventeen times as powerful as the original Daventry transmitter of the BBC.

At that time there had only been two major epidemiological studies according to the NRPB report. This is not strictly true. Although the New York State Power Lines Project reports and the English version of the Polish survey of Szmigielski (which found a sevenfold morbidity ratio for hemato-lymphatic cancers among military personnel exposed occupationally to microwave or radio frequency radiation), were not yet published, the Polish version of his study had actually been published in 1985 (8.34:241). Some of the studies already mentioned, moreover, were large enough to rate inclusion in a document of the kind prepared for the public enquiry. The NRPB cites the large-scale study of 40,000 American personnel of whom half 'may have been rather more exposed to radar emissions during the nineteen fifties (the Korean War) than the other 20,000 who were the control or reference group. . .and as with many studies the degree of exposure could not be established with any certainty' (8.35:216).

Since this study was so imprecise that it inevitably entangled other factors, and in any case did not cover radio frequencies, one might have expected the NRPB not to weaken their own credibility by relying on it. The more so since there were other more precise epidemiological studies pointing in the opposite direction; Allen Frey had been carefully researching low intensity RF hazards since the sixties, and a potted version of this research, published by Andy Marino in *Modern Bioelectricity* in 1988, shows the pioneering work he did (8.36:91).

Frey's own work unearthed the hazard that RF frequency can induce cardiac arrhythmia and even cessation, if the RF pulse coincided with the QRS complex. Anecdotal evidence indicated that cats, rats and rhesus monkeys all attempt to avoid this radiation. Others in the sixties found that RF radiation resulted in a significant decrease in the human auditory threshold. After the passage of Public Law 90-602 in the United States in 1969 research began anew into the bioeffects of EM energy.

The radio frequency research of the seventies is nevertheless largely confined to cellular or experimental studies, investigations into 'RF hearing' (a phenomenon which has grown to afflict thousands today, and is known as 'the Hum'), brain chemistry, and the permeability of the blood brain barrier (BBB). In all these areas the influence of RF radiation was detected. Any hopes that these somewhat disturbing biological findings would be translated into full-scale epidemiological studies were disappointed. As Frey states:

The significant research, that which does not use high intensities and is not thermoregulatory oriented, began to be tapered off about 1980. Though such research received only a very small fraction of the huge amount that has been spent (several hundred million dollars) on RF bioeffects research since 1970, it

has largely been squeezed out for reasons unrelated to science
(8.37:91).

Anyone who wishes better to understand those reasons might
care to read Nicholas Hildyard's book *Cover Up*. In the section on microwaves he mentions the 1964 Johns Hopkins study
linking Down's Syndrome — the fatal radiation-induced chromosome break which causes Mongolism — with the exposure
of fathers to radar; the attempted silencing of Milton Zaret; and
the abandonment of Dietrich Beischer's work on triglycerides.
He also says:

Since the Innsworth enquiry Dr Hawkins has been trying to find
funds in order to conduct a long-term survey of ionization at
Fishpond and elsewhere. Although at one point he was offered
£10,000 by the Health and Safety Executive to undertake a
survey, that offer was subsequently withdrawn (8.38:127).

Szmigielski reviews the progress of bioeffects of EM energy, this
time concentrating on the immune system. His veritable
treasure-house of Eastern Bloc reports, incidentally, includes
a significant paper by Jakovleva in 1974, showing that electromagnetic fields cause a reduction in antibodies to salmonella
(8.39:137). Battery hen and egg laying methods involve 24-hour
irradiation of chicken and eggs alike by electric heat and light,
and this may well be what is causing and will continue to cause,
the salmonella outbreaks in the United Kingdom and the United
States.
   Szmigielski also takes a look at the Korean study mentioned
by the NRPB, saying:

The relevant and acceptable study on delayed health effects in
US navy personnel exposed to radar during the Korean war
should be mentioned here. No significant differences were found
by Robinette et al., and Silverman, between the high and low
exposure groups for malignant neoplasms as the causes of
hospitalization and/or death (from records of Navy and VA
hospitals). However, when three subgroups of the high-exposure
group were developed to provide a gradient of potential exposure,
a trend appeared for increased numbers of malignant neoplasms
in the subgroup rated as highly exposed. The weak point in this
analysis is the fact that only subjects hospitalized in Navy or VA
hospitals were analysed, and only during a certain time. These

subjects were only a part of the total population of United States military personnel that operated in Korea during 1950-1954, and it is difficult, for example, to evaluate which part of the total population was hospitalized or died 1950-1974 (8.40:241).

So much for the Korean study, which if anything seems to support a correlation between high exposure to RF fields and malignancy, rather than the reverse.

Modern Bioelectricity is unlikely to be read by many: it costs £150 at retail prices! However a number of other brave populist writers like Hildyard have attempted to draw the issue to public attention over the succeeding years with their articles and books.

British television by contrast, possibly because it disseminates the very medium by electromagnetic means, has limited its attention to one or two programmes confined for the most part to power frequency and ionizing radiation hazards, such as *Inside Britain's Bomb, The Good the Bad and the Indefensible,* and *The Killing Fields.* One can sympathize with the dilemma in which they find themselves; it must be painful for the managers of such an immediate medium to be unable to address themselves to one of the most important social issues of this millenium, and one which is going to be a major environmental issue during the next decade.

The NRPB's presentation to the public enquiry also referred to a study by Dr Dennis concerning the irradiation of the American Embassy in Moscow:

An exhaustive comparison of all symptoms, conditions, diseases, and causes of death among the employees and their dependents was unable to establish differences in health status by any measure that could be attributed to the electromagnetic fields (8.41:75).

This is in stark contrast to Becker's account in *The Body Electric*:

In January 1977, the State Department under duress announced results of a series of blood tests on returning embassy personnel: a 'slightly higher than average' white blood cell count in about a third of the Moscow staff. If 40 per cent above the white blood cell counts of other foreign service employees (levels common to incipient leukaemia) can be considered 'slightly higher than average', then this technically wasn't a lie. The finding has been officially ascribed to some unknown microbe.

As part of Project Pandora in the late 1960s the State Department tested its Moscow employees for genetic damage upon their return stateside, telling them the inner cheek scrapings were to screen for those unusual bacteria. No results were ever released, and they're reportedly part of the missing files, but one of the physicians who conducted the tests was quoted by the Associated Press as saying they'd found 'lots of chromosome breaks'. Chromosome breaks can cause Down's Syndrome too.

In 1987 the State Department gave its Moscow employees a 20 per cent hardship allowance for serving in 'an unhealthful post', and installed aluminium window screens to protect the staff from radiation a hundred times weaker than that near many radar stations (8.42:20).

The NRPB submission to the enquiry then reviews registry studies, including that of Sam Milham, concluding (to my puzzlement) that radio and television repairmen are more likely to be exposed to power frequencies than to radio frequencies. While acknowledging that clusters of leukaemia have been associated with nuclear power stations, the NRPB report does not seem to recognize that such plants are also *de facto* high sources of EM energy, by virtue of their emerging power lines. It is not explained why power frequency questions have been introduced into an enquiry concerning a radio frequency mast.

In considering cellular studies the NRPB report is on less comfortable ground, since cellular and laboratory studies showing RF effects mount into the hundreds, and most of the recent work has been at pains to exclude thermal factors. The NRPB simply reports that its own (unreplicated) laboratory study was not able to induce chromosomal changes of human cells *in vitro* as a result of radiation when the temperature was maintained below 40°C. *In vitro* studies of that kind have often been criticized as inferior to *in vivo* studies; what may not happen in the test tube could very well happen in the living body, and CMR theory would accord with this view. Whatever their drawbacks, if the NRPB couldn't find any obvious changes in cellular structure, then others could. To quote from Szmigielski again:

Numerous studies have shown an increased mitogenic response of human blood lymphocytes in culture after temperature elevation above 37°C. Human lymphocytes cultured at 38-40°C respond with faster and increased blastoid transformation (8.43:241).

Of course this isn't chromosomal breaking, but the marked inhibition of lymphocytic cytoxicity which is observed is another way of saying that the immune system is being broken down by microwave irradiation. As Szmigielski puts it:

An overview of the available literature and of our own findings suggests the existence of a biphasic reaction of the immune system to MW/RF radiations — stimulation of the whole system (mainly of humoral immunity) after a single or a few days exposure, followed by gradual, but transient, suppression of the whole immunity with prolongation of the exposure period (up to several months) and/or increasing power density of the fields. Stimulation and suppression of immunity in MW/RF-exposed animals both seem to be transient and inconsistent phenomena. At low power densities the system recovers soon after the exposure. Thermal effects and the concomitant stress situation also stimulate numerous immune functions and should be viewed as a beneficial factor with potential therapeutic applications.

Carcinogenesis has been indisputedly linked causally with ionizing radiation. Microwaves and RF are not quite energetic enough to ionize atoms, that is to strip them of some of their electrons, and most researchers have been content to limit their claims to cancer promotion rather than initiation, once the cancer itself has started (I do not necessarily agree with this). The NRPB report claims that all noted effects of EM fields on organic life are at levels high enough to induce an element of heating. But there is a growing list of good solid research showing effects without any heating. The proposed limit suggested by the NRPB is an absorbed dose limit of 0.4 Watts per kilogram. There are however some doubts about the reliability of this measure; Arthur Guy (mentioned in the NRPB report) recently concluded: 'The whole body average SAR is not a reliable predictor of the effect of microwaves. The response may depend on the localized pattern of energy in the body' (8.44:111).

The same issue of the medical journal which carried Guy's statement also published a study by Dutta and his team, which concluded that 'cell lines derived from tumours of the CNS, neuroblastomas, respond to modulated RF fields in a manner identical to the normal forebrain tissue preparations from new-born chicks and from felines (cats),' and their study was using only about one-tenth (0.05 Watts per kilogram) of the limit suggested by the NRPB (8.45:84).

The Office of Technology Assessment, Russia, Poland, Canada, and Australia are quite specific in saying that hazards from non-ionizing EM energy are not to be dismissed lightly, that much more research is needed, and as the results emerge they cast increasing doubt on the permitted exposure limits currently in force (8.45a:249). From electrically-operated fork-lift trucks to plastic welding machines, from VDUs to bedside mains-driven clock radios, from fluorescent lighting tubes to small-boat radar, long-term biological effects are definitely present.

The research programme necessary to combat these individual hazards is huge, and will stretch well into the next century. In the short term the only advice is: don't expose yourself chronically to their influence.

We have come very far, in terms of technological progress, during a few short decades of the twentieth century. But Nature rarely bestows a riskless benefit, and the pleasures of our unbridled use of electromagnetic energy may soon cloy if its long-term effect is found to be birth defects in our young, incapacity, leukaemias, cancers, and a biologically-damaged genetic future. The one thing we cannot afford to do is to ignore the risk.

Even now, as I sit here at the latest 386 series computer, its low-radiation screen doubly protected by an anti-radiation visor, the little Cereus Peruvianus cactus at hand, a large quartz crystal nearby, and a neutralizer of unknown benefit round my neck, pulsing out a steady 8Hz carrier signal for the alleged good of my lymphocytes, I still do not honestly know whether the vast collection of organic cells which is me is adequately protected. My nagging suspicions are not entirely discounted by the NRPB. In considering athermal effects in their newly published guidelines on EM exposure they say:

There does appear to be evidence for athermal biological effects, particularly of magnetic fields, at all levels of biological organization (8.46:188).

To support this they quote a 1986 report by America's National

Council on Radiation and Measurements (which also in fact appeared in their 1988 enquiry evidence), together with one of the NYPLP papers. Still not entirely convinced, the paper continues:

However, the experimental evidence is often statistically weak and proves difficult to reproduce. It is not possible to say with certainty or quantitatively whether this evidence has any implications for human health. The epidemiological evidence suggests that if the risks of occupational exposures to EM fields are real then they are within the range regarded as tolerable, and should not unduly concern individuals. Nevertheless, the apparent low risks to rather ill-defined populations may conceal higher risks to particular groups or from particular modalities of exposure. The Board regards it as important that basic research and epidemiological studies are continued to determine whether the risks are real, and if so their underlying causality.

And finally:

The Board will issue further advice as the results and conclusions of such research and studies become available, and intends to publish a review of existing biological and epidemiological evidence in the near future.

The world is badly in need of that review, and more importantly, of that research.

References

1   Richard J. ABLIN (5.25)
    'Transglutaminase: Co-factor in the Aetiology of AIDS?'
    Lancet 1:813, (6 April 1985).

2   Richard J. ABLIN (5.25)
    'AIDS: A possible explanation'
    Fed. Proc. 43:782, 1984.

3   Albert ABRAMS (1.8; 2.54; 5.2)
    New Concepts in Diagnosis and Treatment
    Philopolis Press, San Francisco (1916).

4   E.D. ACHESON (6.24; 6.26)
    'Encephalomyelitis associated with poliomyelitis virus'
    Lancet 2:1044-1048, (20 Nov. 1954).

5   Jad ADAMS (1.12; 5.31)
    The HIV Myth
    MacMillan, London (1989), pp. 12-14.

6   W. Ross ADEY (7.19)
    'The cellular environment and signalling through cell
    membranes' in *Electromagnetic Fields and Neurobehavioural
    Function*, O'Connor and Lovely, Liss, New York (1988).

7   E.D. ADRIAN & B.C.H. MATTHEWS (2.12)
    'The interpretation of potential waves in the cortex'
    J. Physiol. 81:440-471 (1934).

8   E.D. ADRIAN & YAMAGUVA (2.8; 3.13)
    'The origin of the Berger Rhythm'
    Brain 58: 323-351 (1935).

9   AIDS UK (2.1; 5.29)
    HEA & PHLSAC (April 1989), London, April 1989.

10   Bruce ALBERTS, Dennis BRAY et al. (2.46; 2.49; 8.13)
     Molecular Biology of the Cell
     Garland Publishing, N.Y. (1983).

11   M. ALTER (2.40)
     'Clues to the cause based on the epidemiology of Multiple Sclerosis' in E.J. Field, *Multiple Sclerosis, A Critical Conspectus*, MTP Press (1977).

12   E.W. ANDERSON (1.2; 2.6)
     Animals as Navigators
     Bodley Head, London (1983).

13   P. ANDERSON & S.A. ANDERSON (2.16)
     'Physiological Basis of the Alpha Rhythm'
     Appleton-Century Crofts, N.Y. (1968)

14   Kaethe BACHLER (6.20)
     Earth Radiation: The startling discoveries of a dowser
     Wordmasters, Manchester (1989).

15   S. BARANSKI & P. CZERSKI (8.7; 8.8)
     Biological Effects of Microwaves
     Dowden Hutchingson and Ross, Stroudsberg, Pa. (1976).

16   M. BARNOTHY & J. BARNOTHY (5.37)
     Biological Effects of Magnetic Fields
     Medical Physics 3. Yearbook Pubs., Chicago (1960).

17   Sir JAMES BARR (1.9)
     'Many Inventions'
     BMJ (20th May 1922), p. 819.

18   S.M. BAWIN & W. ROSS ADEY (7.19; 8.12)
     'Sensitivity of calcium binding in cerebral tissue to weak environmental electric fields oscillating at low frequency'
     Proc. Nat. Acad. Sci. 73: 1999-2003 and *Ann. Nat. Acad. Sci.* 247:74 (1976).

19   Robert BECK (1.16; 4.24)
     'Extremely low frequency magnetic fields and EEG entrainment, a psychotronic warfare possibility?'
     Bio-Medical Research Associates, L.A. (1978).

20   Robert O. BECKER & Gary SELDON (1.14; 1.15; 5.9; 5.20; 8.42)
     The Body Electric: Electromagnetism and the Foundation of Life
     Morrow, N.Y., (1985).

21 Robert O. BECKER & Andrew MARINO (2.38)
Electromagnetism and Life
SUNY Press, Albany, N.Y. (1982).

22 Dietrich BEISCHER & J. BREHL (7.11)
'Search for effects of 45 Hz. magnetic fields on live triglycerides in mice'
Naval Aerospace Res. Lab. NAMR 1197, AD7814 (Jan 1985).

23 Valerie BERAL, S. EVANS et al. (1.21; 6.11; 6.14)
'Malignant melanoma and exposure to fluorescent lighting at work'
Lancet 2. pp. 290-293 (7 August 1982).

24 Hans BERGER (2.7a)
Uber das Elekbrenkephalogram des menschen
First Report. *Arch. Fur Psychiatr. & Nervenkrankheit* 87:527-570 (1929).

25 Hans BERGER (2.9)
Psyche
Gustav Fischer, Jena (1940).

26 Rosalie BERTELL (4.1)
No Immediate Danger
Women's Press, London (1985).

27 Simon BEST & N. KOLLERSTROM (6.8)
Planting by the Moon
Foulsham, London (1980).

28 E. BINDMAN & O. LIPPOLD (2.29; 3.19)
The Neurophysiology of the Cerebral Cortex
Edward Arnold, London pp. 64-84, (1981).

29 Christopher BIRD (3.31a)
Divining
MacDonald and James, London & Sydney (1980).

30 Sir Douglas BLACK (4.8; 4.17; 5.35)
Investigations of the possible increased incidence of cancer in West Cumbria
HMSO, London (1984).

31 C.F. BLACKMAN, S.G. BENANE et al. (8.12)
'Effects of ELF (1-120 Hz.) and modulated 50 Hz. RF fields on the efflux of calcium ions from brain tissues in vitro'
Bioelectromagnetics 6: 1-11 (1985).

32   Carl F. BLACKMAN (8.12)
      'Stimulation of brain tissue in vitro by ELF low intensity
      sinusoidal electromagnetic fields' in
      Electromagnetic Waves and Neurobehavioural Function,
      O'Connor and Lovely, Liss, N.Y. (1988).

33   B.R. BLOOM (2.42)
      'Immunological Changes in Multiple Sclerosis'
      Nature, 287: 275-276 (1980).

34   BRITISH NUCLEAR FUELS PLC (7.2a)
      Nuclear Energy: Don't be left in the dark
      BNF PLC, Warrington, 1988.

35   Paul BRODEUR (1.11; 1.13; 4.4; 8.4)
      The Zapping of America
      W.W. Norton, N.Y. (1977).

36   T.H. BULLOCK (2.22)
      'The trigger of sensitivity in certain fish to electric signals'
      Neurosciences Res. Prog. 15: 17 (1977).

37   Maurice BURTON (2.20)
      The Sixth Sense of Animals
      Dent and Sons, London (1973).

38   P.S. CALLAHAN (2.18)
      'Intermediate and far infrared sensing of nocturnal insects'
      Annals Entomol. Soc. Amer. 58(5): 727-745, (1965).

39   Neil CARLSON (3.3)
      Physiological Behaviour
      Allyn and B (1985).

40   Rachel CARSON (4.3)
      Silent Spring
      Hamish Hamilton, London (1963).

41   CASSELL (1.6)
      Cassell's 'Work' Handbooks
      'Wireless Telegraphy and How to Make the Apparatus'
      Cassell & Co, London (1912).

42   E. CELAN, D. GRADINARU et al. (2.52)
      'The evidence of selective radiation emitted by a cell cul-
      ture which destructively affects some tumoural cell lines' in
      B. Jezowska, B. Kochel et al., *World Scientific*, Singapore,
      pp. 219-225 (1986).

43 CENTERS FOR DISEASE CONTROL (5.28)
Weekly Morbidity and Mortality Reports
Atlanta, Georgia.

44 I. CHANARIN, M. BROZOVIC et al., (5.36)
Blood and its Diseases
Churchill Livingstone, Edinburgh (2nd ed. (1980).

45 J.M. CHARCOT (2.44)
Leçons sur les Maladies du Système Nerveux
Delahaye, Paris (1872).

46 CHASE & MORALES (3.18)
'Suppression of motor and sensory action during PS sleep'
in
McGinty *Brain Mechanisms of Sleep*, Raven, N.Y. (1985).

47 Margaret CHENEY (1.7)
Tesla: Man out of Time
Prentice Hall, London (1981).

48 L.G. CLARK & F.A. HARDING (2.31)
'Comparison of M- and D- twins with respect to some features of the EEG'
Proc. Electrophys. Technol. Ass. 16: 94.

49 Robert W. CLARK (2.62)
Albert Einstein
World Publishing, New York, (1971)

50 Jane G. CLARKE (2.41)
Multiple Sclerosis: A New Theory Concerning Cause and Cure
New Age Science Press, London, (1983).

51 R.W. COGHILL (3.21)
'The electric railway children: an electromagnetic aetiology for cot death?'
Hospital Equipment and Supplies (June 1989) p.9.

52 R.W. COGHILL (5.28)
AIDS: the electromagnetic connection
Jnl. Alt. & Comp. Med. (May 1989).

53 COHEN & BARONDES (2.33)
'Effects of ACCH on learning and memory'
Nature 218: 277-273 (1968).

54   M. COLEMAN, J. BELL et al. (4.23; 8.16)
     'Leukaemia incidence in electric workers'
     Lancet 3: 982-983, (1983).

55   Michel COLEMAN & Valerie BERAL (4.2; 4.22)
     'A review of epidemiological studies of the health effects
     of living near or working with electricity generation and
     transmission equipment'
     Intl. Jnl. Epidemiol. 17(1): 1-13 (1988).

56   M.L. COLONNIER (2.30)
     'The structural design of the neocortex' in
     Brain and Conscious Experience (ed. J.C. Eccles), Springer,
     (1966) p. 18.

57   COMARE (Committee on Medical Aspects of Radiation
     in the Environment). First Report (4.8)
     'The implications of the new data on the release from Sel-
     lafield in the 1950s for the conclusions of the report on
     the investigations of the possible increased incidence of
     cancer in West Cumbria' (Clair H. Bobrow)
     HMSO, London (1986).

58   COMARE Second Report (4.9; 8.19)
     'Investigation of the possible increased incidence of leu-
     kaemia in young people near the Dounreay nuclear estab-
     lishment, Caithness, Scotland'
     HMSO, London (1988).

59   COMARE Third report (4.10)
     'Report on the incidence of childhood cancer in West Berk-
     shire and North Hampshire area'
     HMSO, London (1989).

60   Steven CONNOR & Sharon KINGMAN (2.59;5.34)
     Search for the Virus
     Penguin, London (1989).

61   P. CONTI, G.E. GIGANTE et al. (5.38)
     'Reduced mitogenic stimulation of human lymphocytes
     by extremely low frequency electromagnetic fields'
     FEBS 162(1): 156-160 (1983).

62   P.J. COOK-MAZAFFARI, F.L. ASHWOOD et al (4.10a)
     Cancer incidence and mortality in the vicinity of nuclear
     installations, England and Wales, 1959-1980.
     HMSO (OPCS No. 51), (1987).

63   H. COTTON (2.16)
     Electrical Technology
     Pitman, London, (1967).

64   George Washington CRILE (2.34)
     The Phenomena of Life: a radio-electric interpretation
     Heinemann, U.S. (1936).

65   A.H. CRISP, E. STONEHILL et al. (3.16)
     'Insomnia observed in cases of anorexia nervosa'
     Post Grad. Med. Jnl. 47: 207-213.

66   Manfred CURRY (3.30)
     Curry-Netz
     Herold-Verlag Dr Wetzel, Munich (1978).

67   Manfred CURRY (3.31)
     Bioklimatik
     Bioklimatik Research Institute, Ridereau (1946).

68   *Daily Telegraph*, London (1.2)
     'Rare Bird Spotted'
     15 October, 1988.

69   J. DANGUIR & S. NICOLAIDIS (3.15)
     'Energy metabolism is connected to PS sleep' in
     McGinty, *Brain Mechanisms of Sleep*, Raven, N.Y. (1985).

70   Sarah C. DARBY & Richard DOLL (5.38; 8.20)
     'Fallout, radiation doses near Dounreay, and childhood
     leukaemia'
     BMJ: 294: 607 (1987).

71   Heiner DAUS, G. SCHWARZE et al., (5.26)
     'Reduced CD4+ count, infections, and immune throm-
     bocytopenia without HIV infection'
     Lancet (2). 559-560 (2nd Sept. 1989).

72   A.N. DAVISON (2.45; 3.20)
     'The biochemistry of the myelin sheath'
     Myelination, Thomas, Springfield Ill (1970).

73   Bob DeMATTEO (5.17; 7.2; 7.8; 7.14)
     Terminal Shock: the health hazards of Video Display Terminals
     NC Press, Toronto (1986).

74   William C. DELMENT, S. GREENBERG & R. KLEIN (3.8)
      'The persistence of the REM deprivation effect'
      Assoc. for the Psychophysiological Study of Sleep Washington, U.S.A (1965).

75   J.A. DENNIS (8.33; 8.41)
      'Health issues in the siting of a low frequency transmission mast'
      HMSO: National Radiological Protection Board, R 222 (April 1988).

76   V. DIGERNES & E.G. ASTRUP (6.12a)
      'Are Datascreen terminals a source of increased PCB concentrations in the working atmosphere?'
      Intl. Arch. Occup. Envir. Health 49: 193-197 (1982).

77   M.J. DILLON (6.22)
      'Epidemic neuromyasthenia at the Hospital for Sick Children'
      Post Grad. Med. Jnl. 54 (637): 726-720.

78   M.J. DILLON, W.C. MARSHALL et al. (6.23)
      'Epidemic neuromyasthenia: outbreak at a children's hospital'
      Br. Med. Jnl. 1:301-305 (1974).

79   R.R. DRUCKER-COLIN & C.W. SPANIS (2.58)
      'Protein synthesis increases during deep sleep'
      Progress in Neurobiology 6: 1-12(1976).

80   T.D. DUANE & Thomas BEHRENDT (2.14; 4.25)
      'Extrasensory electroencephalographic induction between identical twins'
      Science (15 Oct. 1965).

81   P.H. DUESBERG (5.24)
      'Retroviruses as carcinogens and pathogens: expectations and reality'
      Cancer Research 47: 1199-1220 (1987).

82   Yuri DUMANSKY, V.M. POPOVICH et al. (7.12)
      'Influence of a low frequency EM field (50Hz) on the functional condition of humans' (in Russian)
      Hygiene and Health (Gig i san) 12(1977).

83   J.D. DUMANSKY & M.G. SHANDALA (7.12)
'The biologic action and hygienic significance of EM fields
of SuperHigh and UltraHigh frequencies in densely popu-
lated areas', in
Biologic Effects and Health Hazards of Microwave Radiation
(P. Czerski et al., eds.) Warsaw (1974).

84   S.K. DUTTA, B. GHOSH et al. (8.45)
'Radiofrequency radiation induced calcium ion efflux
enhancement from human and other neuroblastoma cells
in culture'
BEMS 10: 197-202 (1989).

85   Egon ECKERT (3.25)
'Plotzlicher und unerwarteter Tod in Kleinskindesalter und
elektromagnetische Felder'
Med. Klin. 71: 1500-1505 (37) (1976).

86   ELECTRICITY COUNCIL (1.17; 4.16)
Handbook, London (1988).

87   J.L. ELWOOD, C. WILLIAMSON et al (6.12b)
Malignant melanoma in relation to moles pigmentation,
and exposure to fluorescent and other lighting sources.
Br. Jnl. Cancer 53: 65-74 (1986).

88   Robert ENDROS & K.E. LOTZ (3.34; 6.27)
'Zur Frage der Mikrowellendurchlassigkeit bei Baue-
lementen'
Baumeister-Zeitung H. 11, 12.

89   Barry FOX (1.18a)
'If tube trains affect trees, what do they do to us?'
New Scientist (11 August, 1977).

90   Jean de la FOYE (3.37)
Ondes de Vie, Ondes de Mort
R. Laffont, Paris (1980).

91   Allen H. FREY (8.36; 8.37)
'Evolution and results of biological research with low inten-
sity nonionizing radiation'
Modern Bioelectricity (ed. A. Marino) (1988) pp. 785-837.

92   A.H. FREY (7.20)
'Auditory system response to radio frequency energy'
Aerosp. Med. 32: 1140-1142, (1961).

93   A.H. FREY (7.21)
      'Human auditory system response to modulated elec-
      tromagnetic energy'
      J. Appl. Physiol. 17: 689-692 (1962).

94   A.H. FREY & S.R. FELD (3.22a)
      'Avoidance by rats of illumination with low-power non-
      ionizing electromagnetic energy'
      J. Comp. Physiol. Psychol. 89:183 (1975).

95   Herbert FROHLICH (5.16)
      'The biological effects of microwaves and related questions'
      Adv. in Electron. and Electron Phys. 53: 85-152 (1980).

96   J.P. FULTON. S. COBB et al. (4.14)
      'Electrical wiring configurations and childhood leukaemia
      in Rhode Island'
      Am. Jnl. Epidemiol. 111: 192 (1980).

97   D.R. GADSDON & J.L. EMERY (3.25)
      'Fatty change in the brain in perinatal and unexpected
      death'
      Arch. Dis. in Childhood 51: 42-48 (1976).

97a   M.J. GARDNER, M.P. SNEE et al. (8.25)
      'Results of case control study of leukaemia and lymphoma
      among young people'
      BMJ 300: 423-9 (1990).

98   J.A. GENGERELLI & N.J. HOLTER (8.10)
      'Experiments on stimulation of nerves by alternating elec-
      trical fields'
      Proc. Soc. Exp. Biol. Med. 46: 532 (1941).

99   A.G. GILLIAM (6.20a)
      'Epidemiological study. . . among the personnel of the
      L.A. General Hospital'
      Public Health Bull. No. 240., U.S. Public Health Svce., April
      1938.

100   John GOFFMAN (7.0)
      Radiation and Human Health
      Sierra Books, U.S. (1981).

101   Pierre GLOOR (2.8)
      'Hans Berger on the Electroencephalogram of Man'
      Electroencephalo. & Clin. Neurophys. Suppl. Elsevier (1969).

102 M.K. GOLDHABER, M.R. POLEN et al. (7.27)
'The risk of miscarriage and birth defects among women
who use Visual Display Terminals during pregnancy'
Amer. Jnl. Ind. Med. 13: 695-706 (1988).

103 J. GOLDING, S. LIMERICK, & A. MACFARLANE (3.23)
'Sudden Infant Death: Patterns, Puzzles & Problems
Open Books Publishing, Somerset (1985).

104 E.M. GOODMAN, B. GREENEBAUM et al. (4.12)
'Effects of ELF fields on growth and differentiation of phys-
arum polycephalum'
Radiat. Res. 66: 531-540 (1976).

105 Rolf GORDON (DOHM) (3.29; 6.28)
Are You Sleeping in a Safe Place?
Dulwich Health Society, London (1987).

106 D.C. GREEN (2.28; 2.47)
Radio and Line Transmission B
Pitman, London (1971).

107 W. GREY WALTER (2.11)
The Living Brain
Pelican Books, Middx. (1961).

108 John GRIBBIN (3.12)
In Search of the Double Helix
Corgi, London (1985).

109 Tim GRIFFYD JONES (6.30)
Statement to Press
Univ. Bristol Veterinary School (1989)

110 J. P. GUPTA (5.21a)
'Microwave radiation hazards from radars and other high
power microwave generators'
Defence Science Jnl. 238: 287-292 (July 1988).

111 Arthur GUY (7.15; 7.25; 8.44)
'Paper on exposure to low intensity microwaves'
Ann. Mtg. BEMS, Atlanta, GA (1984).

112 Alexander GURWITSCH (2.50)
Uber den Begriff des embryonalen Feldes
Archiv. Fuer Entwicklung mechanik 51:383-415 (1922)

113 E. HAIDER & I. OSWALD (3.17)
'Protein Synthesis enhanced during Paradoxical sleep'
BMJ 2: 318-322 (1970).

114 Virgil K. HANCOCK (5.43)
'Treatment of blood stream infections with haemo-
irradiation'
Amer. Jnl. Surgery 58(3): 336-344 (1942).

115 H.A. HANSSON (3.26)
Lamellar bodies in Purkinje cells experimentally induced
by electric fields.
Brain Res. 216: 1-10 (1981a).

116 HANSARD (3.22)
Parliamentary proceedings
20th March 1989.

117 Ernst HARTMANN (3.5)
The Biology of Dreaming
Charles C. Thomas, Springfield (1967).

118 Ernst HARTMANN & Joseph WUST (3.28)
'Uber physicalische Nachweismethoden der sogenamten
'Erdstrahlen'
Geopathie G Beikeft zur Zietschrift Erfahrungsheilkunde',
Ulm (1954).

119 S.M. HARVEY (7.22)
'Characteristics of low frequency electrostatic and elec-
tromagnetic fields produced by visual display terminals'
Ontario Hydro Research Divn. Report No. 82-528 IX (1982).

120 Alfred HAVILAND (3.28)
'The geographical distribution of disease in Great Britain'
Lancet (25 Feb. 1988) p. 365.

121 L.H. HAWKINS (6.3)
'The influence of air ions, temperature, and humidity on
selective well being and comfort'
Jnl. Environ. Psychol. 1: 279-292 (1981).

122 L.H. HAWKINS & D. D'AURIA (4.11)
'Leukaemia risks near nuclear sites'
BMJ 295: 1488 (1987).

123 HEALTH & SAFETY EXECUTIVE (7.24; 7.26)
Working with VDUs
HMSO, London, (1983).

124 Donald HEBB (3.9)
Organization of Behaviour
John Wiley, N.Y. (1949).

125 J.H. HELLER & A.A. TEIXEIRO-PINTO (7.7a)
'A new physical method of creating chromosomal aber-
rations'
Nature 183: 905-906 (1959).

126 Nancy HICKS, Matthew ZACK et al. (5.8)
'Childhood cancer and occupational radiation exposure
in parents'
Cancer 53: 1636-1643 (1984).

127 Nicholas HILDYARD (8.38)
Cover Up: The facts they don't want you to know
New English Library, London (1981).

128 HIPPOCRATES' WRITINGS (2.2)
(ed. Betty Radice)
Penguin, London (1983).

129 A.M. HO, A.C. FRASER-SMITH et al. (1.18)
'Large amplitude ULF magnetic fields produced by a rapid
transit system: close range measurements'
Radio Science, 14: 1011, (1979).

130 D. HO et al. (2.58)
'AIDS and the brain'
Trends in Neurosciences, 9: 91 (1986).

131 J.G. HOFFMAN (2.5)
The Life and Death of Cells
Hutchinson, London (1958).

132 F.C. HOLLOWS & J.C. DOUGLAS (8.28)
'Microwave cataracts in radiolinemen and controls'
Lancet (18 Aug, 1984).

133 R.E. HOPE-SIMPSON (1.4)
'Relationship of influenza pandemics to sunspot cycles'
(M. Kingsbourn, W. Lynn Smith, eds.)
Charles C. Thomas, Springfield, Illinois (1974).

134 Sir Thomas HORDER (1.10)
'Preliminary communication to RSM, 16th January 1925'
Bale & Sons, Daniellson, London (1925).

135 L.E. HOUGHTON & E.I. JONES (6.25)
     Earliest report of ME in U.K.
     Lancet, 1: 196 (1942).

136 Ursula HUWS (1.19; 4.28; 7.1; 7.18)
     Visual Display Unit Hazards
     London Hazard Centre (1988).

137 M.E. JAKOVLEVA (8.39)
     'Physiological mechanisms of action and electromagnetic
     fields'
     Izd. Medicina, Moscow (1973) (in Russian).

138 Japanese General Council of Trade Unions
     'Japanese miscarriages blamed on computer terminals'
     New Scientist (23 May, 1985).

139 A.A. JENSEN (6.13)
     'Melanoma, fluorescent lights and polychlorinated
     biphenyls'
     Lancet (23 Oct. 1982) p. 935.

140 Don R. JUSTESEN (7.6)
     'Microwave irradiation and blood brain barrier'
     Proc. IEEE vol. 68 i: (Jan. 1980).

141 Bengt KALLEN & Ulrich MORITZ (8.21; 8.24)
     'Delivery outcome among physiotherapists in Sweden: is
     non-ionizing radiation a foetal hazard?'
     Arch. Environ. Health 37 (2): 81-85 (1982).

142 A.J. KALMIJN (2.7)
     'Electroperception in sharks and rays'
     Nature 212: 1232-1233 (1966).

143 V.P. KAZNACHEEV, S.P. SHURIN et al. (2.53)
     'Distant intercellular interactions in a system of two tis-
     sue cultures'
     Psychoenergetic Systems,1: 141-142 (1976).

144 A.R. KENNEDY, M.A. RITTER et al. (6.13a)
     'Fluorescent light induces malignant transformation in
     mouse embryo cell cultures'
     Science 207:1209-1211 (1980)

145 Nathaniel KLEITMAN (3.6)
     Sleep and Wakefulness
     Univ. Chicago Press, Chicago (1939).

146 G.G. KNICKERBOCKER et al. (8.2)
'Study in USSR of medical effects of electric fields of power systems'
Power Engineers Society, Spec. Pub. 10, IEEE (1975).

147 H.L. KONIG, A.P. KRUEGER et al. (6.29)
Biologic effects of environmental electromagnetism
Springer-Verlag, New York (1981).

148 I.H. KORNBLUEH (6.4)
'Brief Review of the Effects of Artificial Ionization of the Air'
Arch. Med. Hydro, 211(3): 1-9,(1961).

149 A.P. KRUEGER & E.J. REED (6.6)
'Biological impact of small air ions'
Science 193: 1209-12 (1976).

150 Thomas S. KUHN (2.61)
The Structure of Scientific Revolutions
Phoenix-University of Chicago Press (1962).

151 Georges LAKHOVSKY (2.19; 3.27; 3.35)
The Secret of Life
W. Heinemann, London (1939).

152 Ioana LANCRANJAN, Madelaine MAICANESCU et al. (8.3)
'Gonadic function in workmen with long-term exposure to microwaves'
Health Physics, 29: 381-383 (1975).

153 R.J. LARKIN & P.J. SUTHERLAND (1.1)
'Migrating birds respond to Project Seafarer's electromagnetic fields'
Science, 195: 777-78 (1987).

154 Peter LAURIE (5.12; 5.14)
Beneath the City Streets
Granada Books (2nd ed.) (1979).

155 P. LEMESURIER (5.46; 6.31)
The Great Pyramid Decoded
Element Books, Dorset (1977).

156 John LESTER & Dennis MOORE (5.41; 7.17)
'Cancer incidence and electromagnetic radiation'
J. Bioelect. 1: 59 (1982).

157 R. LIBURDY (5.5)
'RF radiation alters the immune system: Modulation of T- and B- lymphocyte levels and cell-mediated immunocompetence by hyperthermic radiation'
Radiation Research, 77: 34-36 (1979).

158 James C. LIN, M.J. OTTENBREIT et al. (5.40)
'Microwave effects on granulocyte and macrophage precursor cells of mice in vitro'
Rad. Res. 80: 292-302 (1979).

159 Ruey S. LIN, Patricia C. DISCHINGER et al. (8.31)
'Brain tumours in electrical workers'
Biological effects of electropollution: brain tumours and exptl. models.
(ed. S.K. Dutta, Information Ventures Inc. (1986).)

160 LONDON HAZARDS CENTRE (1.22; 6.12)
Fluorescent Lighting, a Hazard Overhead
LHC Trust, London (1987).

161 Karl E. LOTZ (6.1; 6.19)
Do You Want To Live Heathily?
Paffrath-Druck KG Remscheid, W. Germany (1982).

162 A.R. LURIA (3.2)
The Working Brain, an Introduction to Neuropsychology
Penguin, London (1973).

163 D.B. LYLE, R.D. AYOTTE et al. (4.13; 5.39)
'Suppression of T-Lymphocyte cytotoxicity following exposure to 60 Hz sinusoidal electric fields'
BEMS 9: 303-313, (1988).

164 Brian MADDOCK & John MALE (6.15)
'Power-line fields and people'
Phys. Bull. 38: 345-347 (1987).

165 J.C. MALE, W.T. NORRIS et al. (8.23)
'Exposure of people to power frequency electric and magnetic fields'
Proc. 23rd. Hanford Life Sciences Symp. Oct. 1984 (NTIS) Conf. 841041).

166 Eugene MARAIS (2.20)
The Soul of the White Ant
Penguin, London (1973).

167 A.A. MARINO (8.35)
Modern Bioelectricity
Dekker, N.Y. (1988).

168 Ronald MARKS (5.4)
The Sun and Your Skin
Macdonald Optima, London (1988).

169 Bryan MATTHEWS (2.43)
Multiple Sclerosis, The Facts
OUP, Oxford (1978).

170 Jacques la MAYA (3.36)
La Medecine de l'Habitation
Dangles, France (1984).

171 M.E. McDOWALL (4.20)
'Mortality of persons resident in the vicinity of electricity transmission facilities'
Brit. Jnl. Cancer 53: 271-9 (1986).

172 M.E. McDOWALL (4.21; 8.15)
'Leukaemia mortality in electrical workers in England and Wales'
Lancet, i 246, (1983).

173 D. McFARLAND (ed.) (2.23)
Oxford Companion to Animal Behaviour (shoals' spaces round predators)
OUP, Oxford (1981).

174 D.J. McGINTY et al. (3.15;3.18)
Brain Mechanisms of Sleep
Raven Press, N.Y. (1985).

175 D.I. McREE & H. WACHTEL (8.11)
'The effects of microwave radiation on the vitality of isolated frog sciatic nerves'
Radiat. Res. 82: 536-546 (1980).

176 B. MERZ (7.28)
Statement, October 1986
Institut de Recherches en Geobiologie, 176, Chardonne.

177 George MILEY & P.M. DUNNING (5.44)
'Ultraviolet Blood irradiation in acute virus-type infections'
Rev. Gastroenterol 15(4): 271.

178   George MILEY (5.45)
      'Treatment of eight cases of atypical pneumonia by
      ultraviolet blood irradiation'
      Amer. Bacter. Soc. (Penn Chapter) (1943).

179   Kjell Hansson MILD & P.A. OBERG (8.6)
      'Neurophysiological effects of electromagnetic fields; a crit-
      ical review'
      Kyoto Symposia (EEG suppl. No. 36): 715-729 (1982).

180   S. MILHAM Jr. (5.42; 8.17)
      'Mortality from leukemia in workers exposed to electric
      and magnetic fields'
      New Eng, Jnl. Med. 302: 249 (1982).

181   S. MILHAM Jr. (8.18)
      'Silent Keys: leukaemia mortality in amateur radio
      operators'
      Lancet, 1: 812 (1985).

182   Baruch MODAN (5.19)
      'Exposure to electromagnetic fields and brain malignancy:
      a newly discovered menace?'
      Amer. Jnl. Indl. Med. 13, 635-637 (1988).

183   F. MOORE (2.26)
      'Geomagnetic disturbance and the orientation of noctur-
      nally migrating birds'
      Science, 196: 682-4 (1977).

184   C.T. MORGAN & E. STELLAR (2.10;3.7)
      Physiological Psychology
      McGraw-Hill, N.Y. (1950).

185   MOUNIER JOUVET (3.16a)
      'Neurophysiology of the states of sleep'
      Physiol.Revs. 47: 117 (1967).

186   A. MYERS, R.A. CARTWRIGHT et al. (4.19)
      'Overhead powerlines and cancer'
      Conference on EM fields in medicine and biology:
      (IEEE Conf. Pub. no. 257, pp. 126-130) (London, 1985).

187   Peter NATHAN (2.3)
      The Nervous System
      OUP, Oxford (1983).

188 NATIONAL RADIATION PROTECTION BOARD (8.34; 8.46)
'Guidance on standards'
HMSO, London (May 1989).

189 J. NAUTA (3.10)
'The role of the hypothalamus in sleep'
Jnl. Neurophysiol. 9: 285-316 (1946).

190 S. NORDSTROM, E. BIRKE et al. (5.6; 8.22)
'Reproductive hazards among workers at high voltage sub-stations'
BEMS 4: 91-101 (1983).

191 P.A. OBERG (8.9)
'Magnetic stimulation of nerve tissue'
Med. Biol. Engineering 11: 55-64.(1973).

192 Mary O'CONNOR & Richard LOVELY (7.19;8.12)
Electromagnetic fields and neurobehavioural function
Alan R. Liss N.Y. (1988).

193 H. OLDFIELD & R.W. COGHILL (2.48)
The Dark Side of the Brain
Element Books, Dorset (1988).

194 John O'NEILL (1.5)
Tesla, Prodigal Genius
Neville Spearman, London (1968).

195 Robert ORNSTEIN & David SOBEL (3.1)
The Healing Brain
Simon & Schuster, N.Y. (1987).

196 I. OSWALD (3.17)
'Human brain protein, drugs, and dreams'
Nature, 223: 893-897 (1969).

197 H. PALM (6.18)
Das Gesunde Haus
Ordo-verlag, Bedensee (1975).

198 Jacob D. PAZ, R. MILLIKEN et al. (8.30)
'Potential ocular damage from microwave exposure during electrosurgery: dosimetric survey'
J. Occup. Med. 292 (7): 580-583 (1987).

199 W.E. PENFIELD & E. BOLDREY (2.32)
     'Somatic and sensory representation in the cerebral cor-
     tex of man as studied by electrical stimulation'
     Brain, 60: 389-443 (1937).

200 F.S. PERRY, M. REICHMANIS et al. (4.27)
     'Environmental power frequency magnetic fields and
     suicide'
     Health Physics 41: 267-277 (1981).

201 F.S. PERRY & L. PEARL (6.16)
     'Health effects of ELF fields and illness in multistorey
     blocks'
     Public Health 102: 11-18, (1988).

202 S. PERRY, L. PEARL et al., (6.17)
     'Power frequency magnetic fields; depressive illness and
     myocardial infarction'
     Public Health 103: 177-180, (1989).

203 PETERS & JONES (3.14)
     The Cerebral Cortex, Vol. 3
     Plenum Press, N.Y. (1985).

204 H. PICKLES (7.29)
     Promotional Leaflet
     Ordosan Ltd., Roundhay, Leeds (undated).

205 J.L. PHILLIPS (5.15)
     'In vitro exposure to electromagnetic fields; change in
     tumor cell properties'
     Intern. Jnl. Rad. Biol. 49: 463-9 (1986).

206 Anthony PINCHING (5.32)
     'A Flat Earth Society manifesto'
     Independent, (17th April 1989)

207 Freiherr Gustav von POHL (3.33)
     *Earth Currents–Causative Factor of Cancer and Other
     Diseases*
     Freich-Verlag, Feucht, (1983).

208 Fritz Albert POPP (2.51)
     'On the coherence of ultraweak photon emission from liv-
     ing tissues'
     E.W. Kilmister (ed.) *Disequilibrium and Self Organization*,
     D. Reidel (1986).

209 Susan PRAUNITZ & Charles SUSSKIND (7.16)
'Effects of chronic microwave irradiation on mice'
IRE trans. on Biomedical Electronics 9: 104 (1962).

210 R. RAMANATHAN, S. CHANDRA et al. (3.24)
'Sudden infant death syndrome and water beds'
NEJM 23-6 (1988) p. 1700.

211 Melvin RAMSAY (6.21)
Post Viral Fatigue Syndrome: The Saga of the Royal Free Disease
Gower Medical Pubs., London (1989).

212 Joan RENDELL (5.13)
Letter
Health Now (June 1989).

213 N.V. REVNOVA (4.13a;5.20a;8.2)
Influence on the organism of high voltage power frequency EM fields in hygiene, occupation and biological effects of RF EM fields, Moscow 1968.

214 Guyon RICHARDS (2.35)
The Chain of Life
John Bale and Daniellson, London, (1934).

215 F.H. ROBINSON (5.1)
'Growth of the Industry'
BBC Handbook, London, (1928).

216 C.D. ROBINETTE, C. SILVERMAN et al. (8.35)
'Effects upon health of occupational exposure to microwave radiation'
Am. J. Epidemiol. 112: 39 (1980).

217 Elaine RON, Baruch MODAN et al., (5.18)
'Mortality after radiotherapy for ringworm of the scalp'
Amer. J. Epidem. 127(4): 713-725 (1988)

217a A.H. ROSE (2.37)
Chemical Microbiology
Butterworth, London (1976)

218 Lucien ROUJON (3.38)
L'Energie Micro-Vibratoire et la Vie
Le Rocher, Paris (1987).

219   D. SAVITZ (4.18a)
      'Childhood cancer and electromagnetic field exposure'
      Am. Jnl. Epidemiol. 128: 21-38 (1988).

220   D. SAVITZ & Eugenia E. CALLE (4.18b)
      'Leukaemia and occupational exposure to electromagnetic
      fields: review of epidemiological surveys'
      J. Occup. Med. 29(1): 47-51 (1987).

221   A.M. SERDUK & E.A. SERDUK (5.21)
      'Electromagnetic Fields: Influence on Population Health
      Condition'
      BEMS, Ann. Mtg, Tucson, (1989).

222   R.K. SEVERSON, R.G. STEVENS et al. (5.39)
      'Acute nonlymphocytic leukaemia and residential exposure
      to power frequency fields'
      Amer. J. Epidemiol. 126(1): 10-20 (1988).

223   Hari SHARMA (7.23)
      'An investigation of a cluster of adverse pregnancy out-
      comes and other health related problems among
      employees working with visual display terminals in the
      accounting offices at Surrey Memorial Hospital'
      Vancouver. (1984).

224   Rupert SHELDRAKE (4.26; 6.9)
      A New Science of Life
      Granada, London (1981).

225   Rupert SHELDRAKE (2.6)
      The Presence of the Past
      Collins, London (1988).

226   M. SIEKIERZYNSKI, P. CZERSKI et al. (8.5)
      'Health surveillance of personnel occupationally exposed
      to microwaves. 2: functional disturbances'
      Aerospace Med. 45(10): 1143-1145 (1974).

227   A.T. SIGLER et al. (5.7)
      'Radiation exposure in parents of children with Down's
      Syndrome'
      Johns Hopkins Med. Jnl. 117: 374-399 (1965).

228   B. SIGURDSSON, T. SIGURJONSSON et al (6.20a)
      'A disease epidemic in Iceland simulating poliomyelitis'
      Amer. J. Hyg. 52:222 (1950).

229 Louis SLESIN (5.22)
'Can cables cause cancer?'
Technology Review (Oct. 1987).

230 C.W. SMITH & E. AARHOLT (2.57)
'Possible effects of environmentally simulated endogenous opiates'
Health Physics, 43(6): 929-930 (1982).

231 Cyril SMITH & Simon BEST (4.30)
Electromagnetic Man
J.M. Dent, Clapham (1989).

232 H. SNOW (2.55)
Cancer and the Cancer Process
I & A Churchill, London (1983).

233 Fred SOYKA with Alan EDMONDS (6.2; 6.7)
The Ion Effect
Bantam, N.Y., (1978).

234 Margaret R. SPITZ & Christine C. JOHNSON (8.19)
'Neuroblastoma and paternal occupation: a case-control analysis'
Am. J. Epidemiol. 121: 924-929 (1985).

235 JOSEPH STANGLE (3.33a)
'Radiation measurements over underground aquifers, (in German).
Bohrtechnik Brunnenbau, Rohr leitungs bau
Nr. 11 (1960).

236 Gerald STERN & Andrew LEES (2.14)
Parkinson's Disease, the Facts
OUP, Oxford (1982).

237 F.G. SULMAN (6.5)
'Influence of artificial air ionization on the human EEG'
Int. J. Biometeoral. 18: (1974).

238 F.G. SULMAN (6.5a)
The effect of air ionization, electric fields, atmospherics, and other electric phenomena on man and animal
C.C. Thomas, Springfield, Illinois (1980).

239 J. SURBECK (7.30)
'Protection against noxiousness generated by computer cathodic screens'
Presented at *BEMS* Ann. Mtg. Tucson, (1989).

240 M. SWICORD et al. (7.3; 7.9)
Presentation at Ann. Mtg. Amer. Phys. Soc.,
Detroit (March 1984).

241 Stanislaw SZMIGIELSKI, Marian BIELEC et al. (5.23; 8.33;
8.40; 8.43)
'Immunologic and cancer-related aspects of exposure to
low-level microwave and radio frequency fields' in
Modern Bioelectricity (Marino), Dekker, N.Y., (1988) pp.
861-925.

242 R.H. TAYLOR, R.A.F. COX et al. (4.11)
Leukaemia risks near nuclear sites.
Letter, *BMJ* 296: 213-214 (1988).

243 R.A. TELL & E.D. MANTIPLY (5.11)
'Population exposure to VHF and UHF broadcast radia-
tion in the United States'
Proc. IEEE, 68(1):6-13, (1980).

244 T.S. TENFORDE (2.21)
'Electroreception and magnetoreception in simple and
complex organisms'
BEMS, 10(3): 215-221, (1989).

245 *The* TIMES (1.3; 2.22; 2.25; 3.32)
(on lost pigeons)
London (28 June 1988).

246 *The* TIMES (8.1)
Obituary: Walter Stoessel
(12 December, 1986).

247 L. TOMENIUS (4.18)
'50 Hz electromagnetic environment and the incidence of
childhood tumours in Stockholm County'
BEMS 7: 191-207 (1986).

248 U.S. Air Force (5.10)
'Analysis of the exposure levels and potential biologic
effects of the PAVE-PAWS Radar system'
Environmental Impact Statement, N.A.S., (May 1979) p. 81.

249 US CONGRESS OFFICE OF TECHNOLOGY ASSESS-
     MENT (8.45a)
     Biological effects of power frequency electric and magnetic
     fields.
     Background paper OTA BP-E53
     Washington D.C. May 1989.

250 U.S. Navy (5.8a)
     Report on Radar Hazards
     Medical Bulletin, July 1943.

251 D. VAGERO & R. OLIN (8.14)
     'Incidence of cancer in the electronics industry: using the
     new Swedish cancer environment registry as a screening
     instrument'
     Br. J. Ind. Med. 40: 188-192 (1983).

252 Kristoffer VALERIE, Anne DELERS et al. (5.27)
     'Activation of HIV type 1 by DNA damage in human cells'
     Nature, 333: 78-81 (1988).

253 R.N. VYVYAN (5.3)
     Marconi and Wireless
     EP Publishing, Yorkshire (1974).

254 C. WALCOTT, J.L. GOULD et al. (1.3)
     'Pigeons have magnets'
     Science, 205: 1027-8 (1979).

255 Graham WALKER (2.27)
     Sheffield Star (suggests microwave origin for lost pigeons)
     (July 1988).

256 M.M. WALKER & M.E. BITTERMAN (2.17)
     'Conditioning analysis of magnetoreception in honey bees'
     BEMS 10(3): 261-275 (1989).

257 George de la WARR and Douglas BAKER (2.36)
     Biomagnetism
     De la Warr Labs, Oxford, (1967).

258 J.D. WATSON & F.H.C. CRICK (2.13)
     'Molecular structure of nucleic acids: A structure for deox-
     yribonucleic acid'
     Nature 171:737-738 (1953).

259 Lyall WATSON (3.11)
     The Biology of Death
     Sceptre Books, Hodder & Stoughton (1987).

260 S.J. WEBB (7.5)
     'Synthesis of DNA by specific frequencies'
     Ann. N.Y. Acad. Sci. 247:327 (1975).

261 S.J. WEBB & A.D. BOOTH (7.5; 7.10)
     'Absorption of microwaves by organisms'
     Nature 222: 1268-69 (1969).

262 Robert WECHSLER (2.56)
     'A new prescription: mind over body'
     Discover (Feb. 1986).

263 N. WERTHEIMER & E. LEEPER (4.7; 4.12)
     'Electrical wiring configurations and childhood cancer'
     Am. Jnl. Epidemiol. 109: 273-84 (1979).

264 N. WERTHEIMER & E. LEEPER (4.15)
     Reply to Fulton et al. (1980)
     Am. Jnl. Epidemiol. 111 (4): 461-2 (1980).

265 Morton WHEELER (7.4)
     Microwave News, 4(6) (July 1984).

266 Thomas WILLIS (2.4)
     Cerebri Anatome
     London, 1664

267 E.O. WILSON (2.24)
     The Social Insects
     Harvard Press, Cambridge, Mass (1971).

268 Marlys H. WITTS et al. (5.30)
     'A case of AIDS in 1968'
     JAMA: 251 (20): (25 May 1984).

269 Valerie WORWOOD (6.10)
     Aromantics
     Pan, London (1987).

270 L.B. YOUNG & H.P. YOUNG (4.6)
     'Pollution by electrical transmission'
     Bull. Atom. Scientist 34-38 (Dec. 1974).

271 L.B. YOUNG et al. (4.5)
     Power over People
     OUP, Oxford (1974).

272 M.M. ZARET (7.13)
'Analysis of factors of laser radiation producing retinal damage'
Fed. Proc. Suppl., 14: p. 62-64, (1965).

273 Milton M. ZARET (8.26)
'Cataracts and Visual Display Units' in
Pearce, B.G., *Health Hazards of VDTs*,
Wiley & Sons, Chichester (1984).

274 Milton M. ZARET (8.27)
'Statement for Subcommittee on Investigation and Oversight'
U.S. Committee on Science & Technol.
(12 May, 1981).

275 M.M. ZARET & Wendy Z. SNYDER (8.29)
'Cataracts in aviation environments'
Lancet (26 Feb. 1977) pp.484-485.

276 N.A. ZHURAKOVSKAYA (4.29)
'Effect of low-intensity, high-frequency EM energy on the cardiovascular system'
JPRS L5615, (1976) p. 13.

Bibliography

To those wishing to read more about electropollution or to understand better the basis of organic life are recommended the following books:

What is life? Erwin Schrodinger, (Cambridge University Press); a physicist's view of how life might work, and the essential prelude to our understanding of the physics of life itself.
The Body Electric, Robert Becker (Morrow, N.Y.: 1985); a clearly-written account of decades working to uncover the mechanisms of wound repair and protein synthesis in electromagnetic terms.
Rupert Sheldrake's *A New Science of Life* (Granada, 1981) and *The Presence of the Past* (Collins, 1988) have had a formative influence not only on me but also on many scientists tackling the problems of morphology.
The Secret Life of Plants, by Peter Tompkins and Christopher Bird (Harper & Row, 1984) also touches on the electromagnetism of life with a well-written account of the earlier pioneers of these researches. An enchanting introduction to the mysteries of our planet's life forms.
The Biology of Death, by Lyall Watson (Sceptre, 1987) (previously titled *The Romeo Error*), will open new avenues for readers interested in the mechanics of life.
Electromagnetic Man, by Simon Best and Dr Cyril Smith (Dent, 1989); a detailed account of the scientific arguments raging about electropollution, containing a plethora of detailed research.

In choosing these few I am most reluctantly omitting many others which in their own way are equally important or enjoyable. Some represented milestones in their time. You will find most of them listed in the references.

Index